Collins

easy l
How to use English

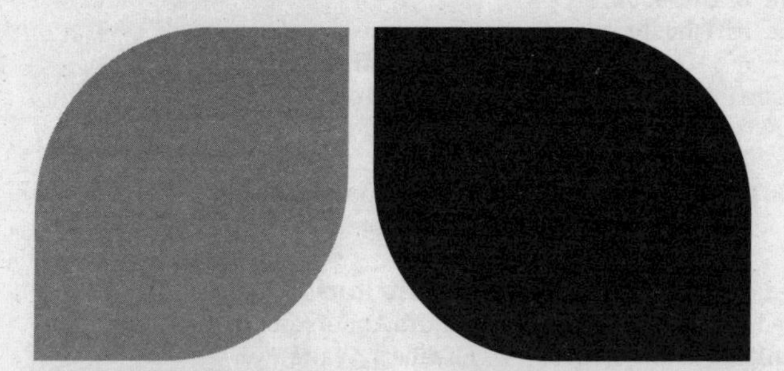

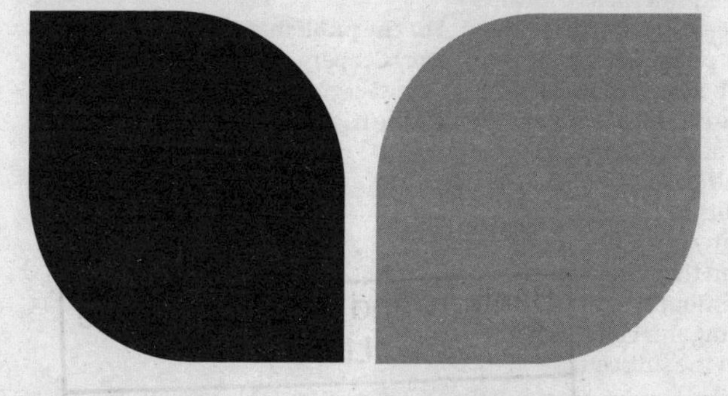

Everythin ow
for natur nglish

HarperCollins Publishers
Westerhill Road
Bishopbriggs
Glasgow
G64 2QT

First edition 2011

10 9 8 7 6 5 4

ISBN 978-0-00-737470-0

www.collinsdictionary.com/cobuild
www.collinselt.com

A catalogue record for this book is available from the British Library

Typeset by Davidson Publishing Solutions, Glasgow

Printed in Great Britain by Clays Ltd, St Ives plc

Senior Editors:
Penny Hands
Kate Wild

Project management:
Lisa Sutherland

Contributors:
Sandra Anderson
Elizabeth Walter

For the publisher:
Lucy Cooper
Kerry Ferguson
Elaine Higgleton

contents

Introduction v

Guide to entries vi - vii

Grammatical terms viii

Pronunciation guide x

How to Use English A-Z **1 - 296**

Topics

Adjectives that cannot be used in front of nouns 297

Times of the day 298

Seasons 300

Transport 300

Meals 301

Places 303

Negatives 304

North, South, East and West 306

Talking about men and women 307

Where you put adverbs 309

Collins Easy Learning How to Use English is designed for anyone who wants to improve their ability to use English accurately. Whether you are preparing for an exam, or you are curious about how a particular word is used, or you simply want a quick guide to common mistakes and how to correct them, *Collins Easy Learning How to Use English* offers you the information you require in a clear and accessible format.

This book focuses on words and structures that often cause problems. Here are some examples of the kinds of problems that are covered:

- words that are easily confused. For example, *hard* and *hardly* look similar, but they have very different meanings.
- words that have similar meanings but are used in different grammatical ways. For example, *allow* and *let* have the same meaning, but we *allow someone to do something*, whereas we *let someone do something*.
- common words that have a lot of different meanings in English, such as *take*, *at* and *better*.
- words that are used with particular prepositions, for example *look for* and *look after*; *made from*, *made of* and *made out of*.
- common 'false friends': English words that look similar to words in other languages, but have different meanings.

You will also find information on which uses are more suitable for formal English (for example, when writing essays) and which ones are more suitable for informal English (for example, when writing emails or in everyday conversation). In addition, there are notes on British English and American English.

Collins Easy Learning How to Use English is arranged alphabetically, so that you can easily find what you are looking for. There are also ten topic sections at the end, where some sets of words and structures are covered together.

For more information on products to help you improve your English, please visit us at **www.collinselt.com**.

Entries are organized in alphabetical order.
accommodation
This information explains clearly and simply what a word or phrase means, and how to use it.
Accommodation is where you live or stay, especially when you are on holiday or when you are staying somewhere for a short amount of time.
Examples from the Collins corpus show how words and phrases are used in real English.
We booked our flights and accommodation three months before our holiday.
There is plenty of student accommodation in London.
Common differences between American and British English are shown.
In British English, **accommodation** is an uncountable noun. In American English, it is usually a countable noun.

Entries that deal with different words, senses or uses are clearly divided into sections.

Special notes give you extra information and highlight common mistakes.

Crossed-out examples show incorrect uses. The correct form is given in a typical example of everyday use.

Cross-references show other entries or topics where you can find more information.

alone – lonely

1 'alone'

If you are **alone**, you are not with any other people.

I wanted to be alone.

Don't use 'alone' in front of a noun. For example, don't talk about '~~an alone woman~~'. Instead, say 'a woman **on her own**'.

These holidays are popular with people on their own.

→ see also topic: **Adjectives that cannot be used in front of nouns**

grammatical terms

adjective: a word that is used for telling you more about a person or thing. You use an adjective to talk about appearance, colour, size, or other qualities, e.g. *A tall woman; She has brown eyes; My gloves are wet*.

adverb: a word that gives more information about when, how, or where something happens, e.g. *See you tomorrow; He spoke slowly; I want to get down*.

auxiliary verb: a verb that is used with another verb in perfect and progressive forms, or to form questions, negatives and passives. The main auxiliary verbs in English are *be*, *have* and *do*, e.g. *Is it snowing?; I have never been to Washington; We don't have a computer at home.* **Modal verbs** are also a kind of auxiliary verb.

clause: a group of words containing a verb. Some sentences contain one clause, e.g. *I fell asleep*. Other sentences contain more than one clause, e.g. *When I told him, he laughed; I want to go but I feel too ill*.

comparative: an adjective or adverb with *-er* on the end or *more* in front of it, e.g. *taller, more beautiful, more carefully*.

conjunction: a word such as *and*, *but*, *if*, and *since*. Conjunctions are used for linking two words or clauses, e.g. *I enjoyed my holiday, but it wasn't long enough; James and Ewan came to the party; If you miss your bus, you'll have to walk home*.

countable noun: a noun used for talking about things that can be counted. Countable nouns have both singular and plural forms, e.g. *This is a lovely house; They are building several new houses*.

determiner: a word such as *a*, *the*, *that* or *my*, which you put in front of a noun to show which person or thing you mean, e.g. *This is his car; Look at that bird*.

future form: a form with *will*, *shall* or *be going to*, which you often use when you are talking about the future, e.g. *He will come soon; I'm going to visit Sarah*.

infinitive: the base form of a verb, which can be used with or without *to*, e.g. *(to) see; (to) bring*.

infinitive without 'to': the infinitive of a verb without *to*, e.g. *Let me think; I must go*.

'-ing' form: a verb that ends in *-ing*, '-ing' forms are used in the progressive, and also after certain verbs, e.g. *I was walking along the beach when I saw him; Please stop shouting*.

intransitive verb: a verb that does not have an object, e.g. *She arrived; I waited*.

linking verb: a verb such as *be*, *become*, *feel* or *seem*. Linking verbs link the subject of a clause with an adjective or a noun phrase that tells you more about the subject, e.g. *I feel sad; She became a doctor*.

modal verb: a verb such as *may*, *must*, or *would*. In a statement, you usually put a modal verb in front of the infinitive form of a verb. Modal verbs do not add an *-s* in the third person singular (with *he*, *she*, and *it*), e.g. *He could win if he tried harder*.

noun: a word that refers to a person, a place, a thing, or a quality, e.g. *Where's Linda?; Go to my room and fetch my bag, please; Unemployment is a problem in London*.

noun phrase: a group of words that functions as a noun. A noun phrase can be a pronoun, a noun, or a noun along with other words such as adjectives, e.g. *She arrived; The old man smiled*.

object: a noun phrase that often shows what is affected by the verb. In most statements, the object follows the verb, e.g. *She ate a sandwich; I locked the door*.

passive form: a form such as *was given* and *were taken*. In the passive, the subject is usually something or someone that is affected by the verb e.g. *Many trees were destroyed; A decision was made by the committee.*

past participle: the form of a verb that is used in perfect forms and passives. Many past participles end in *-ed*, e.g. *talked, jumped, decided*. Others are irregular, e.g. *been, had, given, taken, seen.*

past simple: the past tense form of a verb that is usually used to talk about past events and situations, e.g. *I saw him last night; We talked for hours.*

past tense form: the form of a verb that is used for the past simple. Many past tense forms end in *-ed*, e.g. *talked, jumped, decided*. Others are irregular, e.g. *was/were, had, gave, took, saw.*

perfect form: a form that is made with *have* and a past participle, e.g. *Have* you *seen him?; Someone had eaten all the biscuits.*

phrase: a group of words that are used together and have a meaning of their own, e.g. *The operation was necessary in order to save the baby's life.*

preposition: a word such as *by, with*, or *from* which is always followed by a noun phrase or an '-ing' form, e.g. *He stood near the door; Alice is a friend of mine; This knife is for slicing bread.*

present simple: a form that is often used to talk about habitual actions or permanent states. It is either the base form of a verb, or a verb ending in – *s*, e.g. *I go to work by car; She loves him.*

progressive form: a form that is often used to talk about ongoing situations. It is formed of the verb *be* and an '-ing' form, e.g. *I am enjoying this party; We were having dinner when he phoned.*

pronoun: a word that you use instead of a noun, when you do not need or want to name someone or something directly, e.g. *John took the book and opened it; He rang Mary and invited her to dinner.*

subject: a noun phrase that often shows the person or thing that does the action expressed by the verb. In most statements, the subject comes in front of the verb, e.g. *Tom laughed; The tree fell over.*

superlative: an adjective or adverb with *-est* on the end or *most* in front of it, e.g. *happiest, most intelligent, most carefully.*

'to'-infinitive: the infinitive with *to*, e.g. *I like to drive; She wanted to leave.*

transitive verb: a verb that has both a subject and an object, e.g. *She dropped the mug; We made dinner.*

uncountable noun: a noun that is used for talking about things that are not normally counted, or that we do not think of as single items. Uncountable nouns do not have a plural form, and they are used with a singular verb, e.g. *He shouted for help; We got very wet in the rain; Money is not important.*

verb: a word that is used for saying what someone or something does, or what happens to them, or to give information about them, e.g. *She slept till 10 o'clock in the morning; I ate my breakfast quickly.*

verb phrase: a group of words that functions as a verb. A verb phrase can be a single verb, or it can be a verb along with one or more auxiliary verbs, e.g. *She laughed; We must leave; He could be lying; I'll call you tomorrow.*

A

a – an

You use **a** and **an** when you are talking about a person or thing for the first time. The second time you talk about the same person or thing, you use **the**.

She picked up a book.
The book was lying on the table.

→ see **the**

You can describe someone or something using **a** or **an** with an adjective and a noun.

We live in an old house in the country.

When you say what someone's job is, use **a** or **an** in front of the name of the job. For example, say 'He is **an** architect'. Don't say '~~He is architect~~'.

She became a lawyer.

about

1 'about'

You use **about** when you mention what someone is saying, writing, or thinking.

She told me about her job.
I need to think about that.

2 'about to'

If you are **about to do** something, you are going to do it very soon.

He was about to leave.

Don't use an '-ing' form in sentences like these. Don't say, for example, '~~He was about leaving~~'.

→ see also **around – round – about**

above – over

1 used to describe position

If something is higher than something else, you can use either **above** or **over**.

He opened a cupboard above the sink.
There was a mirror over the fireplace.

2 used to describe amounts and measurements

Above and **over** are both used to talk about a measurement or level of something that is higher than a particular amount.

The temperature rose to over 40 degrees.
...everybody above five feet eight inches in height.

Don't use 'above' in front of a number when you are talking about a quantity or number of things or people. For example, don't say '~~She had above thirty pairs of shoes~~'. Say 'She had **over** thirty pairs of shoes' or 'She had **more than** thirty pairs of shoes'.

It cost over 3 million pounds.
He saw more than 800 children there.

You use **over** to say that a distance or period of time is longer than the one mentioned.

...a height of over twelve thousand feet.
Our relationship lasted for over a year.

accept – except

Don't confuse **accept** /əksept/ with **except** /ɪksept/.

1 'accept'

Accept is a verb. If someone offers something to you and you **accept** it, you agree to take it.

She never accepts presents from clients.

2 'except'

Except is a preposition or conjunction. It is used to show that you are not including a particular thing or person.

All the boys except Patrick started to laugh.

→ see **except**

accommodation

Accommodation is where you live or stay, especially when you are on holiday or when you are staying somewhere for a short amount of time.

We booked our flights and accommodation three months before our holiday.
There is plenty of student accommodation in London.

In British English, **accommodation** is an uncountable noun. In American English, it is usually a countable noun.

The hotel provides cheap accommodations and good food.

according to

You can use '**according to**' when you want to report what someone said.

According to Eva, the train is always late.

You can also use '**according to**' when you want to report the information in a book, newspaper or report.

They drove away in a white van, according to a police report.

> Don't say '~~according to me~~'. If you want to say what your opinion is, you can say '**in my opinion**'.
>
> *In my opinion, all children should learn to swim.*
>
> Also, don't use **according to** and **opinion** together. Don't say, for example, '~~According to Dr. Hussein's opinion, John died of a heart attack~~'. Say '**According to** Dr. Hussein, John died of a heart attack' or 'Dr. Hussein's **opinion** is that John died of a heart attack.'

actually

You can use **actually** when you want to emphasize that something is true, especially if it is surprising or unexpected.

Some people think that Dave is bad-tempered, but he is actually very kind.

You can use **actually** if you want to correct what someone says.

'Lynne was a doctor for ten years.' – 'Eleven years, actually.'

If someone suggests something and you want to suggest something different, you can say '**Actually**, I'd rather ...', or '**Actually**, I'd prefer to ...'.

'Shall we go to the cinema?' – 'Actually, I'd rather go shopping.'

> Don't use 'actually' when you want to say that something is happening now. Use **currently**, **at the moment**, or **now**.
>
> *He's in a meeting at the moment.*

advice – advise

1 'advice'

Advice /ədvais/ is a noun. If you give someone **advice**, you tell them what you think they should do.

She promised to follow his advice.

> **Advice** is an uncountable noun. Don't say '~~advices~~' or '~~an advice~~'. You can say **a piece of advice**.
>
> *Could I give you a piece of advice?*

2 'advise'

Advise /ədvaiz/ is a verb. If you **advise** someone to do something, you say that you think they should do it.

He advised her to see a doctor.

advise

→ see **advice – advise**

affect – effect

1 'affect'

Affect /əfekt/ is a verb. To **affect** someone or something means to cause them to change, often in a negative way.

These problems could affect my work.

2 'effect'

Effect /ɪfe̲kt/ is a noun. An **effect** is something that happens or exists because something else has happened.

They are still feeling the effects of the war.

You can say that something **has an effect on** something else.

Her words had a strange effect on me.

a few

→ see **few – a few**

afford

If you **can afford** something, you have enough money to pay for it. If you **can't afford** something, you do not have enough money to pay for it.

It's too expensive – we can't afford it.
When will we be able to afford a new TV?

You use **afford** with **can**, **could**, or **be able to**. Don't say that someone 'affords' something.

Don't say that something '~~can be afforded~~'. Say that **people can afford** it.

We need to build houses that people can afford.

afraid – frightened

If you are **afraid** or **frightened**, you think that something bad will happen.

The children were so afraid that they ran away.
They felt frightened.

You can also say that you are **afraid of** something or someone, or **frightened of** something or someone.

Tom is afraid of the dark.
Lu was frightened of her father.

> Don't use 'afraid' in front of a noun. For example, don't say '~~an afraid boy~~'. Say 'a **frightened** boy'.
>
> → see also topic: **Adjectives that cannot be used in front of nouns**

If you are worried about something, you can say that you are **afraid of** doing something wrong, or **afraid that** something will happen. You don't usually use 'frightened' in this way.

Keira was afraid of being late for the meeting.
I was afraid that nobody would believe me.

If you have to tell someone something and you think it might upset or annoy them, you can politely say **I'm afraid...**, **I'm afraid so**, or **I'm afraid not**. You can't use 'frightened' in this way.

'Can you remember her name?' – 'I'm afraid not.'
'I'm afraid Sue isn't at her desk at the moment. Can I take a message?'

afternoon → see topic: **Times of the day**

ago

You use **ago** to say how much time has passed since something happened. For example, if it is now 2010 and something happened in 2005, it happened five years **ago**.

We met two years ago.

You use a verb in the past simple with **ago**.

I did it a moment ago.

Don't use **ago** and **since** together. Don't say, for example, '~~It is three years ago since it happened~~'. Say 'It happened **three years ago**' or '**It is three years since** it happened'.

He died two years ago.
It is two weeks since I wrote to you.

agree

If someone says something and you say, **I agree**, you mean that you have the same opinion.

'That film was excellent.' – 'I agree.'

You can also say that you **agree with** someone or **agree with** what they say.

I agree with Mark.
He agreed with my idea.

Don't say that you 'agree' something, or that you 'are agreed with it'. Also, when you use 'agree' in this sense, don't use progressive forms. Don't say '~~I am agreeing with Mark~~'.

If you **agree to do** something, you say that you will do it.

She agreed to lend me her car.

Don't say that you '~~agree doing~~' something.

If people make a decision together, you can say that they **agree on** it.

We had a meeting with our clients and we agreed on a price.

You can also say that people **agree to do** something together.

We agreed to meet at 2 o'clock.

alike

→ see topic: **Adjectives that cannot be used in front of nouns**

a little

→ see **little – a little**

alive

→ see topic: **Adjectives that cannot be used in front of nouns**

allow – let

If you **allow** someone to do something, or **let** someone do something, you give them permission to do it.

1 'allow'

Allow is followed by an object and a 'to'-infinitive.

We allow the children to watch TV after school.

You can say that people **are not allowed to** do something, or that something **is not allowed**.

Visitors are not allowed to take photographs in the museum.
Dogs are not allowed in the gardens.

2 'let'

Let is followed by an object and an infinitive without 'to'.

I love sweets but my dad doesn't let me eat them very often.

> You can't use a passive form of **let**. Don't say, for example, '~~He was let go~~' or '~~He was let to go~~'.

3 'let ... know'

If you **let** someone **know** something, you tell them about it.

I'll let you know what happened.

4 'let me'

You can use **let me** when you are offering to do something for someone.

Let me take your coat.

→ see also **let's – let us**

alone – lonely

1 'alone'

If you are **alone**, you are not with any other people.

I wanted to be alone.

> Don't use 'alone' in front of a noun. For example, don't talk about '~~an alone woman~~'. Instead, say 'a woman **on her own**'.
>
> *These holidays are popular with people on their own.*
>
> → see also topic: **Adjectives that cannot be used in front of nouns**

2 'lonely'

If you are **lonely**, you are unhappy because you do not have any friends or anyone to talk to. **Lonely** is used either in front of a noun or after a linking verb like **be** or **feel**.

He was a lonely little boy.
She must be very lonely here.

already

You use **already** when something happened earlier. Using **already** sometimes suggests that something happened earlier than you expected. Speakers of British English use **already** with a verb in a perfect form. They put **already** after **have**, **has**, or **had**, or at the end of a sentence.

'Would you like some lunch?' – 'No thanks, I've already eaten.'
The train has left already.

Speakers of American English usually use the past simple with 'already'. For example, in British English you say 'I have already seen that film'. In American English you say 'I **already saw** that film' or 'I **saw** that film **already**'.

> Don't confuse **already** with **still** or **yet**. Use **still** when something that existed in the past continued and exists now. Use **yet** when something has not happened, although it probably will happen in the future.
>
> *Donald is 89 and he is still teaching.*
> *They haven't finished yet.*
>
> → see **still**
> → see **yet**

also – too – as well

If you want to add information to something you have said, you use **also**, **too**, or **as well**.

1 'also'

You put **also** after the verb **be**, but in front of all other main verbs.

Rebecca is very clever, and she is also hard-working.
He loves football, and he also plays tennis.

You can put **also** at the beginning of a sentence.

This computer is very modern and fast. Also, it's cheap.

Don't put **also** at the end of a sentence.

2 'too' and 'as well'

You put **too** or **as well** at the end of a sentence.

It's cold outside, and it's raining, too.
Could I have two coffees, please? And a slice of cake as well.

Don't put **too** or **as well** at the beginning of a sentence.

always

1 used to mean 'at all times' or 'forever'

If something **always** happens, it happens at all times. If it has **always** happened, or will **always** happen, it has happened forever or will happen forever.

If there is no auxiliary verb, **always** goes in front of the verb.

Talking to Harold always cheered her up.

If the verb is **be**, you put **always** after it.

She was always in a hurry.

If there is an auxiliary verb, you put **always** after it.

I've always been very careful.

When you use **always** with this meaning, don't use it with a verb in a progressive form. Don't say, for example, '~~Talking to Harold is always cheering her up~~'.

2 used to talk about things that often happen

If you say that something is **always** happening, you mean that it happens often and that it annoys you. When you use **always** like this, you use it with a verb in a progressive form.

Why are you always interrupting me?

among

→ see **between – among**

an

→ see **a – an**

anniversary – birthday

An **anniversary** is a date when you remember something special that happened on that date in an earlier year.

It's our wedding anniversary today.
...the 400th anniversary of Shakespeare's birth.

You don't call the anniversary of the date when you were born your 'anniversary'. You call it your **birthday**.

Mum always sends David a present on his birthday.

another

1 used to mean 'one more'

Another thing or person means one more thing or person of the same kind. **Another** is usually followed by a singular countable noun.

Could I have another cup of coffee?

You can use **another** with a number in front of a plural countable noun.

The woman lived for another ten days.

2 used to mean 'different'

Another thing or person also means a different thing or person.

It all happened in another country.

> Don't use 'another' in front of a plural noun or an uncountable noun. Don't say, for example, '~~They arrange things better in another countries~~'. Say 'They arrange things better in **other** countries'.
>
> *Other people had the same idea.*
> *...toys, paints, books and other equipment.*

anxious

→ see **nervous – anxious – irritated**

any

→ see **some – any**

anybody

→ see **someone – somebody – anyone – anybody**
→ see topic: **Talking about men and women**

anyone

→ see **someone – somebody – anyone – anybody**
→ see topic: **Talking about men and women**

anything

→ see **something – anything**

anyway

1 'anyway'

You use **anyway** when you want to show that something is true despite something else that has been said.

I'm not very good at chess, but I play it anyway.

2 'any way'

Anyway is different from **any way**. You usually use **any way** in the phrase **in any way**, which means 'in any respect' or 'by any means'.

He didn't hurt her in any way.
If I can help in any way, please ask.

anywhere

→ see **somewhere – anywhere**

appreciate

If you **appreciate** something that someone has done for you, you are grateful to them because of it.

Thanks. I really appreciate your help.

You use **appreciate** with **it** and a clause beginning with **if** to say politely that you would like someone to do something.

We would really appreciate it if you could come.

> You must use **it** in sentences like this. Don't say, for example, '~~We would really appreciate if you could come.~~'

argument

→ see **discussion – argument**

around – round – about

1 talking about movement

When you are talking about movement in many different directions, you can use **around**, **round** or **about**.

They were flying around in a small plane.

I spent a couple of hours driving round Richmond.
Police walk about carrying guns.

2 used as a preposition or adverb

Around and **round** are used as prepositions and adverbs, and have the same meaning.

She was wearing a scarf round her head.
The lady turned around angrily.

Around is more common in American English than in British English.

3 meaning 'approximately'

About is used to mean 'approximately'.

He's about forty.

Around and **round about** are also used to mean 'approximately' in conversation.

He owns around 200 acres.
I've been here for round about ten years.

Don't use 'round' like this.

arrange

→ see **manage – arrange**

arrive – reach – get to

You use **arrive**, **reach** and **get to** to say that someone comes to a place at the end of a journey.

I'll tell Professor Hogan you've arrived.
He reached Bath in the late afternoon.
We got to the hospital at midnight.

1 'arrive'

You usually say that someone **arrives at** a place.

We arrived at Victoria Station.

However, you say that someone **arrives in** a country or city.

He arrived in France on Tuesday.

> You never say that someone '~~arrives to~~' a place.
>
> Don't use '**arrive at**' or '**arrive in**' in front of **home**, **here** or **there**.
>
> *I arrived here yesterday.*
> *They arrived home before us.*

2 'reach'

> **Reach** always comes in front of a noun. Don't say, for example, '~~I reached at their house~~' or '~~I reached to their house~~'.
>
> *It was dark by the time I reached their house.*

as … as

1 in comparisons

When you are comparing one person or thing to another, you can use **as** followed by an adjective or adverb followed by another **as**.

You're just as bad as your sister.
I can't run as fast as you can.

After these expressions, you can use either a noun phrase and a verb, or a noun phrase on its own.

You're as old as I am.
Roger is nearly as tall as his father.

Use the pronouns **me**, **him**, **her**, **us** or **them** after **as ... as**.

He looked about as old as me.

However, if the pronoun is followed by a verb, use **I**, **he**, **she**, **we** or **they**.

The teacher is just as happy as they are.

2 used with negatives

You can also use words like **not** and **never** in front of **as** ... **as** to make negative sentences.

Linda is not as clever as Louise.

3 used to describe size or amount

You can use expressions such as **twice**, **three times**, or **half** in front of **as** ... **as** to compare the size or amount of something with something else.

...volcanoes twice as high as Everest.

ashamed – embarrassed

1 'ashamed'

If you are **ashamed**, you feel sorry about something you did wrong.

He upset Dad, and he feels a bit ashamed.

You can say that someone is **ashamed of** something, or **ashamed of** themselves.

Jen feels ashamed of the lies she told.
I was ashamed of myself for getting so angry.

2 'embarrassed'

If you are **embarrassed**, you are worried that people will laugh at you or think you are foolish.

He looked a bit embarrassed when he noticed his mistake.

You can say that someone is **embarrassed by** something or **embarrassed about** it.

I was really embarrassed about singing in public.
He seemed embarrassed by the question.

> Don't use 'of' in sentences like these. Don't say, for example, '~~He seemed embarrassed of the question.~~'

ask

You say that someone **asks** a question, or **asks** someone a question.

The police officer asked me lots of questions.

> Don't use 'to' when you talk about asking someone a question. Don't say, for example, '~~The police officer asked to me a question~~'. Also, don't say that someone '~~says a question~~'.

You also use **ask** when you are reporting questions.

We asked her if she spoke French.
I asked Tom where he lived.

You can use **ask** when you are directly reporting a question.

'How are you?' he asked.

You use **ask for** when you want to report a request. For example, if a man says to a waiter 'Can I have a cup of tea?', you can say 'He

asked for a cup of tea' or 'He **asked** the waiter **for** a cup of tea'.

You can also report a request by saying that someone **asks** someone **to do** something.

He asked me to open the window.

asleep

→ see **sleep**
→ see topic: **Adjectives that cannot be used in front of nouns**

assist – be present

1 'assist'

If you **assist** someone, you help them. **Assist** is a formal word.

We may be able to assist with the tuition fees.

2 'be present'

If you want to say that someone is there when something happens, you say that someone **is present**.

There was no need for me to be present.

as soon as

You use **as soon as** to say that something will happen immediately after something else has happened.

As soon as we get the tickets we'll send them to you.

> Don't use a future form after **as soon as**. Don't say, for example, '~~I will call you as soon as I will get back~~'. Say 'I will call you as soon as I **get** back'.
>
> *Ask him to come in as soon as he arrives.*

When you are talking about the past, you use the past simple or the past perfect form after **as soon as**.

As soon as she got out of bed the telephone stopped ringing.
As soon as she had gone, he started eating the cake.

assume

→ see **suppose – assume**

as well

→ see **also – too – as well**

at

1 place or position

At is used to talk about where something is or where something happens.

There was a staircase at the end of the hall.

You say that someone sits **at** a table or desk, for example when they are eating or writing.

I was sitting at my desk reading.

At is used to talk about the building where something is or where something happens.

We had dinner at a restaurant in Attleborough.

You say that something happens **at** an event such as a meeting or a party.

Mike and Anne first met each other at a dinner party.

→ see also topic: **Places**

2 time

You use **at** to talk about when something happened or will happen.

At 2.30 a.m. he returned.

You say that something happens **at** Christmas or **at** Easter.

She always sends a card at Christmas.

However, you say that something happens **on** a particular day during Christmas or Easter.

They played cricket on Christmas Day.

In British English, **at** is used with **weekend**.

I went home at the weekend.

American speakers usually use **on** or **over** with **weekend**.

I had a class on the weekend.

→ see also topic: **Meals**
→ see also topic: **Times of the day**

attempt → see **try – attempt**

autumn → see topic: **Seasons**

awake → see topic: **Adjectives that cannot be used in front of nouns**

away – far

1 'away'

If you want to state the distance of a particular place from where you are, you say that it is that distance **away**.

Durban is over 300 kilometres away.
The camp is hundreds of miles away from the border.

Don't use 'far' when you are stating a distance. Don't say, for example, '~~Durban is over 300 kilometres far.~~'

2 'far'

You use **how far** when you are asking about a distance.

How far is it to York?

You also use **far** in questions and negative sentences to mean 'a long distance'.

Tell us about your cottage. Is it far?
It isn't far from here.

Don't use 'far' like this in positive sentences. Don't say, for example, that a place is 'far'. Say that it is **far away** or **a long way away**.

The lightning was far away.
His house is quite a long way away from here.

B

bad – badly

1 'bad'

Something that is **bad** is unpleasant or harmful.

When we heard the bad news, we were very upset.
Too much coffee is bad for you.

The comparative and superlative forms of **bad** are *worse* and *worst*.

The storm is getting worse.
It was the worst day of my life.

2 'badly'

> Don't use 'bad' as an adverb. Don't say, for example, '~~I did bad in my exam~~'. Say 'I did **badly** in my exam'.
>
> *The project was badly managed.*

When **badly** is used like this, its comparative and superlative forms are *worse* and *worst*.

Badly has another different meaning. If you need or want something **badly**, you need or want it very much.

I badly need this job.

When **badly** is used like this, its comparative and superlative forms are *more badly* and *most badly*.

badly

→ see **bad – badly**

bag

A **bag** is a container made of paper or plastic that something is sold in.

...a bag of crisps.

A **bag** is also a soft container for carrying things.

Mia put the shopping bags on the kitchen table.

You can call a woman's handbag her **bag**.

Carol took her mobile phone out of her bag.

You can call someone's luggage their **bags**.

They went into their hotel room and unpacked their bags.

A single piece of luggage is a **case** or a **suitcase**.

The taxi driver helped me with my suitcase.

bare – barely

1 'bare'

Bare is an adjective. You can describe a part of the body as **bare** if it is not covered with any clothing.

Jane's feet were bare.

You can also say that a surface is **bare** if it is not covered or decorated with anything.

The flat has bare wooden floors.

2 'barely'

Barely is an adverb. It has a totally different meaning from **bare**. If you can **barely** do something, you can only just do it. If something is **barely** noticeable, you can only just notice it.

She was so afraid, she could barely breathe.
Jawad's whisper was barely audible.

> Don't use 'not' with **barely**. Don't say, for example, '~~We could not barely hear him~~'. Say 'We could **barely** hear him'.

barely

→ see **bare – barely**

bath – bathe

1 'bath'

In British English, a **bath** /bɑːθ/ is a long container that you fill with water and sit or lie in to wash your body.

She was lying in the bath.

In American English, a container like this is called a **bathtub** or a **tub**.

I got into the bathtub.

If you **bath** someone, you wash them in a bath.

We need to bath the baby.

> Don't say that you '~~bath yourself~~'. Say that you **have a bath** or **take a bath.**
>
> *I'm going to have a bath.*
> *In the afternoon she took a bath.*

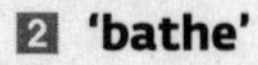

2 'bathe'

In American English, instead of saying that someone has a bath or takes a bath, you can say that they **bathe** /beɪð/.

I went back to my apartment to bathe and change.

bathe

→ see **bath – bathe**

be able to

→ see **can – could – be able to**

bear – can't stand – put up with

1 'bear'

Bear is used to talk about experiencing unpleasant situations. The other forms of **bear** are *bears, bore, borne.*

If you talk about someone **bearing** pain or an unpleasant situation, you mean that they accept it in a brave way.

It was painful, of course, but he bore it.

2 'can't bear'

Bear is often used in negative sentences. If you say that you **can't bear** something or someone, you mean that you dislike them very much.

I can't bear him!

3 'can't stand'

You can also say that you **can't stand** someone or something if you dislike them very much.

He kept on shouting and I couldn't stand it any longer.

4 'put up with'

If you **put up with** something, you accept it, although you do not like it.

The local people have to put up with a lot of tourists.

beat

If you **beat** someone or something, you hit them several times very hard.

His stepfather used to beat him.

The other forms of **beat** are *beats, beat, beaten.*

They beat him, and left him on the ground.

If you **beat** someone in a game, you win the game.

Arsenal beat Oxford United 5-1.

→ see also **win – defeat – beat**

be born

When a baby **is born**, it comes out of its mother's body.

My mother was forty when I was born.

You often say that a person **was born** at a particular time or in a particular place.

Caro was born on April 10th.
Mary was born in Glasgow in 1959.

become – get – go

1 'become'

When a person or thing **becomes** something, they start to be that thing. If you **become** a doctor, a teacher, or a writer, for example, you start to be a doctor, a teacher, or a writer.

Greta wants to become a doctor.

If someone or something **becomes** a certain way, they start to have that quality.

When did you first become interested in politics?

2 'get'

In conversation, **get** is sometimes used to talk about how people or things change and start to have a different quality. It can be followed only by an adjective, not a noun.

I'm getting cold.
It's getting dark.

3 'go'

Go is used to talk about a sudden change in a person's body. Like **get**, it can be used only before an adjective. For example, you can say that someone **goes** blind or deaf.

He went blind twenty years ago.
Katrina went red with embarrassment.

Go is always used in the phrases **go wrong** and **go mad**.

Something has gone wrong with our car.
Tom went mad and started shouting at me.

begin

→ see **start – begin**

behind

1 used as a preposition

If you are **behind** something, you are at the back of it.

They parked the motorcycle behind some bushes.

> Don't use 'of' after **behind**. Don't say, for example, '~~They parked the motorcycle behind of some bushes~~'.

If you are **behind schedule**, you are later doing something than you had planned.

The helicopter was seven minutes behind schedule.

2 used as an adverb

Behind can also be an adverb.

The other police officers followed behind in a second vehicle.
Several customers have fallen behind with their payments.

believe

If you **believe** someone, or if you **believe** what they say, you think that what they say is true.

Please believe me.
I don't believe a word you're saying.

If you **believe** that something is true, you think that it is true.

Police believe that the fire was started deliberately.

If you say that you **believe in** something you mean that you believe it exists.

I don't believe in magic.

You can also say that you **believe in** an idea. This means that you think that it is good or right.

...a country that believes in justice and freedom.

> **Believe** is not used in progressive forms. For example, don't say '~~I am believing you~~'. Say 'I **believe** you'.

belong

1 showing possession

If something **belongs to** you, it is yours.

Everything you see here belongs to me.

> When **belong** is used with this meaning, it must be followed by **to**. Don't say, for example, '~~This bag belongs me~~'. Say 'This bag **belongs to** me'.
>
> **Belong** is not used in progressive forms. For example, don't say '~~This money is belonging to my sister~~'. Say 'This money **belongs to** my sister'.

2 showing where something or someone should be

You can also use **belong** to say that someone or something is in the right place.

The plates don't belong in that cupboard.

be made up of

→ see **comprise – be made up of**

be present

→ see **assist – be present**

better

Better is the comparative form of both 'good' and 'well'. Don't say that something is '~~more good~~' or is done '~~more well~~'. Say that it is **better** or is done **better**.

A written letter is sometimes better than an email.
The team is playing better than ever.

You can also say that someone is **better**, or is feeling **better**.

This means that they are recovering, or that they have recovered, from an illness or injury.

The doctor thinks I will be better by the weekend.

If you say that someone **had better** do something, you mean that they should do it. **Had better** is always followed by an infinitive without 'to'. People usually shorten **had** to **'d**. They say 'I'd better...', 'We'd better...' and 'You'd better...'.

You'd better hurry if you want to get there on time.
We'd better not say anything.

between – among

1 describing position

If something is **between** two things, it has one thing on one side and the other thing on the other side.

Janice was standing between the two men.
Northampton is roughly halfway between London and Birmingham.

> Don't say that something is **between** several things. Say that it is **among** them.
> *There were teenagers sitting among the adults.*

2 differences

You talk about a difference **between** two or more things. Don't use 'among'.

What is the difference between European and American football?
There isn't much difference between the three parties.

3 choosing

You say that someone chooses **between** two or more things. Don't use 'among'.

She had to choose between work and her family.
Choose between tomato, cheese or meat sauce on your pasta.

big – large

When you are describing the size of an object, you can say that it is **big** or **large**. **Big** is usually used in conversation, and **large** is more formal.

This is a big house, isn't it?
Most of the large houses had been made into flats.

1 'large'

Use **large** to describe amounts. You don't usually talk about 'a big amount' or 'a big number'.

A large number of students passed the exam.

2 'big'

Use **big** when you are describing a problem or danger. You don't usually talk about 'large problems'.

Traffic is one of London's biggest problems.

Don't say that a cold or a headache is 'big' or 'large'. Use an adjective such as **bad** or **terrible**.
I have a bad cold.
I've got a terrible headache.

birthday → see **anniversary – birthday**

bit

1 'bit'

A **bit** is a small amount or a small part of something.

There's a bit of cake left.
He found a few bits of wood in the garage.

2 'a bit'

A bit means 'to a small degree'.

She looks a bit like her mother.
He was a bit deaf.

> Don't use **a bit** with an adjective in front of a noun. Don't say, for example, '~~He was a bit deaf man~~'.

3 'a bit' with negatives

You can add **a bit** at the end of a negative statement to make it more strongly negative.

She hadn't changed a bit.

bite

→ see **sting – bite**

blame – fault

If you **blame** someone or something **for** something bad that happened, you think that they made it happen. You can also **blame** something **on** someone.

Police blamed the bus driver for the accident.
Don't blame me!
Jane blames all her problems on her parents.

You can also say that someone is **to blame** for something bad that has happened.

It was an accident – no-one was to blame.

> Don't say that something is someone's 'blame'. Say that it is their **fault**.
>
> *It's not our fault if the machine breaks down.*
> *This was all Rishi's fault.*

blow up

→ see **explode – blow up**

bookshop

→ see **library – bookshop**

border – frontier

1 'border'

The **border** between two countries is the line between them.

They crossed the border into Mexico.

2 'frontier'

A **frontier** is a border with official points for people to cross, often with guards.

They introduced stricter frontier controls.

You talk about one country's border or frontier **with** another.

...a small Dutch town near the border with Germany.

bored – boring

1 'bored'

If you are **bored with** something or someone, you are not interested in them.

I am bored with this film.

If you have nothing to do, you can say that you are **bored**.

Many children get bored during the summer holidays.

2 'boring'

Don't confuse **bored** with **boring**. If you say that something is **boring**, you mean that it is not interesting.

It was a very boring job.
He's a kind man, but he's a bit boring.

boring

→ see **bored – boring**

borrow – lend

If you **borrow** something that belongs to someone else, you use it for some time and then return it.

Could I borrow your pen?
I borrowed this book from the library.

If you **lend** something you own to someone else, you allow them to use it for some time. The past tense form and past participle of **lend** is *lent*.

She lent me £50.
Would you lend me your calculator?

You don't usually talk about borrowing or lending things that cannot move. You ask to 'use' something, or you say that you will 'let someone use' something.

Could I use your garage next week?
She let me use her office while she was on holiday.

brand – make

A **brand** is a product that has its own name, and is made by a particular company. You use **brand** to talk about things that you buy in shops, such as food, drink, and clothes.

This is my favourite brand of cereal.
The advert promotes a new brand of shampoo.

Don't confuse **brand** with **make**. You use **make** to talk about the names of products such as machines or cars.

This is a very popular make of bike.

> Don't talk about the 'mark' of a product. For example, don't say '~~What mark of coffee do you drink?~~' Say 'What **brand** of coffee do you drink?' Don't say '~~What mark of car do you drive?~~' Say 'What **make** of car do you drive?'

You always use '**brand** of' and '**make** of' followed by an uncountable noun or singular noun.

breakfast → see topic: **Meals**

briefly → see **short – shortly – briefly**

bring – take – fetch

1 'bring'

If you **bring** someone or something with you when you come to a place, you have them with you.

Please bring your calculator to every lesson.

The past tense form and past participle of **bring** is *brought*.

My secretary brought my mail to the house.
I've brought you a present.

If you ask someone to **bring** you something, you are asking them to carry it to the place where you are.

Can you bring me some water?

2 'take'

If you **take** someone or something to a place, you carry or drive them there. The past tense form of **take** is *took*. The past participle is *taken*.

He took me to the station.

If you **take** someone or something with you when you go to a place, you have them with you.

Don't forget to take your umbrella.

3 'fetch'

If you **fetch** something, you go to the place where it is and return with it.

I went and fetched another glass.

bring up – raise – educate

1 'bring up'

When you **bring up** children, you look after them throughout their childhood, as their parent or guardian.

Ron was brought up in a working-class family.
When my parents died, my grandparents brought me up.

2 'raise'

Raise can be used to mean 'bring up'.

Lien raised three children on her own.
They want to get married and raise a family.

3 'educate'

Bring up and **raise** do not have the same meaning as **educate**. When children are **educated**, they are taught different subjects over a long period, usually at school.

I was educated in an English public school.

burglar → see **thief – robber – burglar**

business

Business is the work of making, buying, and selling goods or services. In this sense, **business** is an uncountable noun.

Are you going to Tokyo for business or pleasure?

> When you use business in this sense, don't say 'a business'. Don't say, for example, '~~I've got a business to do~~'. Say 'I've got **some business** to do'.

You can talk about a particular area of business as 'the' + noun + '**business**'.

Cindy works in the music business.
My brother is in the restaurant business.

A **business** is a company, a shop, or an organization that makes and sells goods or provides a service. In this sense, **business** is a countable noun.

She owns a successful hairdressing business.

but

You use **but** to introduce something that contrasts with what you have just said.

1 used to link clauses

But is usually used to link clauses.

It was a long walk but it was worth it.

2 used to link adjectives or adverbs

You can also use **but** to link adjectives or adverbs that contrast with each other.

We are poor but happy.
Quickly but silently she ran out of the room.

buy

When you **buy** something, you get it by paying money for it. The past tense form and past participle of **buy** is *bought*.

I'm going to buy everything that I need today.
He bought a first-class ticket.

If you pay for a drink for someone else, you say that you **buy** them a drink.

Let me buy you a drink.

Don't say '~~Let me pay you a drink~~.'

by

1 used in passives

By is most often used in passive sentences. If something is done or caused **by** a person or thing, that person or thing does it or causes it.

She was woken by a loud noise.
I was surprised by the letter.

2 used with time expressions

If something happens **by** a particular time, it happens at or before that time.

I'll be home by eight o'clock.
By 1995 the population had grown to 3 million.

3 used to describe position

You can use **by** to say that something is beside or close to something.

She was sitting in a chair by the window.
...a cottage by the sea.

Don't use 'by' with the names of towns or cities. Use **near** instead.

Winston Churchill was born near Oxford.

4 saying how something is done

You can use **by** with some nouns to say what you use to do something. You don't usually put a determiner (a word such as 'a', 'that' or 'my') in front of the noun.

Are you paying by cash or cheque?
He sent the form by email.
I always go by train.

→ see also topic: **Transport**

However, if you want to say that you use a particular object or tool to do something, you often use **with**, rather than 'by'. **With** is followed by a determiner.

Turn the meat over with a fork.

can – could – be able to

Both **can** and **could** are followed by an infinitive without 'to'.

Can you sing?
I could work for twelve hours a day.

1 negative forms

The negative form of **can** is **cannot** or **can't**. The negative form of **could** is **could not** or **couldn't**. To form the negative of **be able to**, you put **not** or another negative word in front of **able**.

Many people here cannot afford telephones.
They couldn't sleep.
We were not able to give any answers.

2 ability: the present

You use **can** or **be able to** to talk about ability in the present. **Be able to** is more formal than **can**.

You can all read and write.
The sheep are able to move around in the shed.

3 ability: the past

You use **could** or a past form of **be able to** to talk about ability in the past.

He could run faster than anyone else.
I wasn't able to do these quizzes.

4 ability: the future

You use a future form of **be able to** to talk about ability in the future.

I will be able to answer that question tomorrow.

5 awareness

Can and **could** are used with verbs such as **see**, **hear**, and **smell** to say that someone is or was aware of something.

I can smell gas.
I can't see her.
I could see a few stars in the sky.

6 permission

Can and **could** are used to say that someone is or was allowed to do something.

You can have anything you want.
At last the police officer said I could go home.

Cannot or **can't** and **could not** or **couldn't** are used to say that someone is or was not allowed to do something.

You can't bring children into the restaurant.
His mum said he couldn't go to the party.

cancel

→ see **delay – cancel – postpone – put off**

can't stand

→ see **bear – can't stand – put up with**

care

1 'care'

If you **care** about something, you feel that it is important or interesting. You can use **care about** followed by a noun, or **care** followed by a clause beginning with a word like **what**, **who** or **if**.

He really cares about the environment.
I don't care what my dad says about it.

2 'care for' and 'take care of'

If you **care for** people or animals, or you **take care of** them, you look after them.

You must learn how to care for children.
I'll take care of the children.

carry – take

1 'carry'

If you **carry** something to a place, you hold it in your hands and take it there.

He picked up his suitcase and carried it into the bedroom.

2 'take'

If you **take** something to a place, you move it from one place to another. You can move it with your hands, or in other ways, such as in a car.

She gave me some books to take home.
It's his turn to take the children to school.

catch

→ see topic: **Transport**

certainly

→ see **surely – definitely – certainly**

chairman – chairwoman – chairperson

1 'chairman'

The **chairman** is the person who is in charge of a meeting or an organization.

The chairman welcomed us and opened the meeting.
...Andrew Knight, chairman of News International.

2 'chairwoman'

The **chairwoman** is the woman who is in charge of a meeting or an organization.

...Elaine Quigley, chairwoman of the institute.

3 'chairperson' and 'chair'

You can also use **chairperson** or **chair** to talk about either a man or a woman who is in charge of a meeting or an organization.

She is the chairperson of the planning committee.
Please address your remarks to the chair.

chairperson → see **chairman – chairwoman – chairperson**

chairwoman → see **chairman – chairwoman – chairperson**

chance – luck

1 'chance'

If it is possible that something will happen, you can say that there is **a chance that it will happen** or **a chance of it happening**.

There is a chance that I will have to stay longer.
There's no chance of going home.

If something is likely to happen, you can say that there is **a good chance** that it will happen.

We've got a good chance of winning.

If someone is able to do something at a particular time, you can say that they have **the chance to do** it.

You will be given the chance to ask questions.

2 'by chance'

If something happens **by chance**, it was not planned.

Many years later he met her by chance at a dinner party.

3 'luck'

Don't confuse **chance** and **luck**. **Luck** is the good things that happen to you that are not caused by you or by other people.

I couldn't believe my luck.
She hugged me and wished me luck.
Good luck!

→ see also **occasion – opportunity – chance**

cheerful

→ see **glad – happy – cheerful**

chef – chief

1 'chef'

A **chef** /ʃef/ is a cook in a hotel or restaurant.

He worked as a chef in a hotel in Paris.

2 'chief'

The **chief** /tʃiːf/ of a group or organization is its leader.

...the police chief.

chief

→ see **chef – chief**

chips

In British English, **chips** are long, thin pieces of potato fried in oil and eaten hot. Americans call these **fries** or **french fries**.

...fish and chips.
...a steak and fries.

In American English, **chips** or **potato chips** are very thin slices of potato that have been fried until they are hard and crunchy and are eaten cold. British people call these **crisps**.

...a bag of potato chips.
...a packet of crisps.

Christian name

→ see **first name – forename – given name – Christian name**

client

→ see **customer – client**

close – closed – shut

1 verbs

If you **close** /kləʊz/ or **shut** something such as a door, you move it so that it covers or fills a hole or gap. The past tense form and past participle of **shut** is *shut*, not 'shutted'.

Do you mind if I close the window?
He shut the gate.

If a shop or business **closes** or **shuts**, it is not open and people cannot buy or do things there.

The shop closes on Sundays.
What time do the shops shut?

2 adjectives

You can use both **closed** and **shut** as adjectives. You use **shut** after the verb **be**. Don't use it in front of a noun.

The windows were all shut.
The shop was shut all afternoon.

When you are talking about doors and windows, you can use **closed** after **be** or in front of a noun.

The gate was closed.
He could hear voices behind the closed door.

When you are talking about shops and businesses, you put **closed** after **be**.

The supermarket was closed when we got there.

> Don't confuse the verb **close** with the adjective **close** /kləus/. If something is **close** to something else, it is near to it.
>
> *Our house is close to the sea.*

closed

→ see **close – closed – shut**

cloth

→ see **clothes – clothing – cloth**

clothes – clothing – cloth

1 'clothes'

Clothes /kləuðz/ are things you wear, such as shirts, trousers, dresses, and coats.

I took off all my clothes.

> There is no singular form of **clothes**. You can say a **piece of clothing**, but in conversation, you usually name the piece of clothing you are talking about.

2 'clothing'

Clothing /kləuðɪŋ/ is an uncountable noun. You use it to talk about particular types of clothes, for example **winter clothing** or **warm clothing**.

You must wear protective clothing.

3 'cloth'

Cloth /klɒθ/ is an uncountable noun, and means fabric such as wool or cotton which is used for making clothes.

...strips of cotton cloth.

clothing

→ see **clothes – clothing – cloth**

college

→ see topic: **Places**

colour

When you want to describe the colour of something, you usually use a colour adjective such as 'red' or 'green' rather than the word **colour**.

She had green eyes.

> Don't say '~~She had green colour eyes.~~'

However, you use the word **colour** when you are asking about the colour of something.

What colour was the bird?

You can also say that one thing is **the colour of** another thing.

The paint was the colour of grass.

In sentences like these you use **be**, not 'have'. Don't say '~~What colour has the bird?~~' or '~~The paint had the colour of grass~~'.

The American spelling of 'colour' is **color**.

come

1 'come'

You use **come** to talk about movement towards a place.

Mark came to stay with us.
Please come and see me in my office.

2 'come' or 'go'?

When you are talking about movement away from the place where you are, you use **go**, not 'come'.

During the summer we went to France for a week.

→ see **go**

If you invite someone to accompany you somewhere, you usually use **come**, not 'go'.

Will you come with me to the party?

3 'come and'

You use **come and** with another verb to say that someone visits you or moves towards you in order to do something.

Come and see me next time you're in New York.

4 'come from'

If you **come from** a place, you were born there, or it is your home.

'Where do you come from?' – 'Australia.'
I come from Zambia.

Don't use a progressive form in sentences like these. Don't say, for example, '~~Where are you coming from?~~' or '~~I am coming from Zambia.~~'

→ see also **true – come true**

comment – commentary

1 'comment'

A **comment** is something you say that expresses your opinion of something.

People started making rude comments.

2 'commentary'

A **commentary** is a description of an event that is broadcast on radio or television while the event is taking place.

We listened to the football commentary on the radio.

comment – mention – remark

1 'comment'

If you **comment on** a situation, or make a **comment** about it, you give your opinion on it.

Mr Cook has not commented on these reports.
I was wondering whether you had any comments.

If you **mention** something, you say it, but only briefly, especially when you have not talked about it before.

He mentioned that he might go to New York.

If you **remark on** something, or make a **remark** about it, you say what you think or what you have noticed, often in a casual way.

Visitors remark on how well the children look.
They repeated Janet's remarks about Adrienne.

commentary → see **comment – commentary**

compare

1 'compare'

When you **compare** things, you consider how they are different and how they are similar.

Doctors have compared the two treatments.

You can use either **with** or **to** after **compare**.

...studies comparing Russian children with those in Britain.
I haven't got anything to compare it to.

2 'be compared to'

If someone or something **is compared to** or **can be compared to** another person or thing, people say that they are similar.

...a computer virus can be compared to a biological virus.

concentrate

If you **concentrate on** something, you give it all your attention.

Concentrate on your driving.

If you say that someone **is concentrating on** something or **is concentrating on** doing something, you mean that they are spending most of their time or energy on it.

They are concentrating on saving lives.

Don't say that someone 'is concentrated' on something.

consider

If you **consider** something, you think about it carefully.

He had no time to consider the matter.

You can say that someone **is considering doing** something in the future.

They were considering opening an office in Paris.

Don't use a 'to'-infinitive after **consider**. Don't say, for example, '~~They were considering to open an office in Paris.~~'

consist of – be made up of

If you say that one thing **consists of** other things, you mean that those things combine to form it. For example, if a book **consists of** twelve chapters, there are twelve chapters in the book.

The committee consists of scientists and engineers.

You can also say that something **is made up of** other things. This has the same meaning as **consist of**.

All substances are made up of molecules.

Don't use progressive forms of these verbs. Don't say, for example, '~~All substances are being made up of molecules.~~'

constant – continual – continuous

Constant, **continual**, and **continuous** have slightly different meanings. You use **constant** to talk about things that are always there. For example 'constant pain' means pain that does not stop. You use **continual** to talk about things that happen often over a period time, and **continuous** to talk about things that happen without stopping. For example, if you say 'There was continual rain', you mean that it rained often. If you say 'There was continuous rain', you mean that it did not stop raining.

You can put **constant** and **continuous** in front of a noun or after a verb.

...a constant flow of traffic.
The breeze was constant, but not too strong.
The exercise should be one continuous movement.
The noise was almost continuous.

However, **continual** can only be used in front of a noun. Don't use it after a verb.

He was tired of her continual complaining.

content

When **content** is a noun, it is pronounced /kɒntent/. When it is an adjective, it is pronounced /kəntent/.

1 used as a plural noun

The **contents** of something such as a box or room are the things inside it.

She searched through the contents of her handbag.

The **contents** of something such as a document are the things written in it.

He knew by heart the contents of the note.

2 used as an uncountable noun

The **content** of something such as a speech, piece of writing, website, or television programme is the information it gives, or the ideas or opinions expressed in it.

I was shocked by the content of some of the speeches.
The website content includes a weekly newsletter.

3 used as an adjective

If you are **content to do** something, you are happy to do it. If you are **content with** something, you are happy and satisfied with it.

He was content to let her do all the talking.
He was content with his morning's work.

If you are **content**, you are happy and satisfied. When **content** has this meaning, it is not used in front of a noun.

He says his daughter is quite content.

continual → see **constant – continual – continuous**

continuous → see **constant – continual – continuous**

contrary

1 'on the contrary'

You say **on the contrary** when you are saying that the opposite of what has just been said is true.

'You'll hate it.' – 'On the contrary. I'll enjoy it.'

2 'on the other hand'

Don't say 'on the contrary' when you are going to mention something that gives a different opinion from something you have just said. Don't say, for example, '~~I don't like living in the centre of the town. On the contrary, it's useful when you want to buy something~~'. Say 'I don't like living in the centre of the town. **On the other hand**, it's useful when you want to buy something'.

It's a difficult job. But, on the other hand, the salary is good.

control

Control can be a verb or a noun.

1 used as a verb

If someone **controls** something such as a country or an organization, they have the power to take all the important decisions about it.

The Australian government controlled the island.

2 used as a noun

Control is also used as a noun. You say that someone has control **of** a country or organization, or control **over** it.

Mr Ronson gave up control of the company.
The government does not yet have control over the area.

3 another meaning

Control is used in the names of places where your documents and luggage are officially checked, especially when you are travelling between countries.

She went through passport control into the departure lounge.

Don't use 'control' as a verb to mean 'check'. Don't say, for example, '~~My luggage was controlled~~'. Say 'My luggage **was checked**' or 'My luggage **was inspected**'.

convince – persuade

1 'convince'

If you **convince** someone of something, you make them believe it is true.

It took them a few days to convince me that it was possible.

2 'persuade'

If you **persuade** someone to do something, you make them do it by talking to them.

Marsha was trying to persuade Katrina to change her mind.

cook

1 'cook'

If you **cook** something, you prepare it and heat it, for example in an oven or saucepan. **Cook** is only used to talk about food, not drinks.

He cooked a delicious meal.
I need to cook the pasta.

2 'make'

If you **make** a meal or a drink, you combine foods or drinks together to produce something different. You can **make** a meal without heating anything.

I made breakfast for everyone.
I'll make you a coffee.

3 'fix'

In American English, you can use **fix** instead of **make**.

Martin fixed some lunch for us.
Lucinda fixed herself a drink.

There are many verbs that talk about different ways of cooking things:

4 'bake', 'roast'

When you **bake** or **roast** something, you cook it in an oven without liquid. You **bake** bread and cakes, but you **roast** meat.

My husband baked a cake for my birthday.
I roasted the chicken.

5 'boil'

When you **boil** something, you cook it in boiling water.

I'll boil the potatoes.

6 'fry'

When you **fry** something, you cook it in hot fat or oil.

Fry the onions until they are brown.

cooker

A **cooker** is a large piece of equipment in a kitchen that you use for cooking food.

The food was warming in a saucepan on the cooker.

A person who cooks meals is called a **cook**, not a 'cooker'.

Agnes is an excellent cook.

corner

A **corner** is a place where two sides or edges of something meet. You usually say that something is **in** a corner.

...a television set in the corner of the room.

The place where two streets meet is also a **corner**. You use **on** when you are talking about the corner of a street.

There is a post box on the corner.

cost

→ see **price – cost**

could

→ see **can – could – be able to**

country

1 'country'

A **country** is an area of land with its own government.

She'd never lived in an English-speaking country.

2 'the country'

You call land that is away from towns and cities **the country**.

We live in the country.

couple

→ see **pair – couple**

crime

A **crime** is an illegal action. You say that someone **commits** a crime.

A crime has been committed.

> Don't say that someone '~~does a crime~~' or '~~makes a crime~~.'

customer – client

1 'customer'

A **customer** is someone who buys something from a shop or a website.

Mrs Adams is one of our regular customers.

2 'client'

A **client** is a person who pays someone for a service.

A lawyer and his client were sitting at the next table.

D

dare

The verb **dare** has two meanings.

1 used in negative sentences

If someone **daren't** do something, they are not brave enough to do it.

I daren't ring Jeremy again.

In American English, you use the full form **dare not** instead of the short form **daren't**.

I dare not leave you here alone.

If you are talking about the past, you say that someone **did not dare** do something or **didn't dare** do something.

She did not dare leave the path.
I didn't dare speak or move.

2 used as a transitive verb

When **dare** is a transitive verb, it has a different meaning. If you **dare** someone to do something, you invite them to do something dangerous.

They dared me to jump into the water.

day

1 'day'

A **day** is a period of twenty-four hours. There are seven days in a week.

Sheilah left a message a few days ago.

You also use **day** to talk about the time when it is light.

The days were dry and the nights were cold.
The meeting went on all day.

2 'today'

You talk about the actual day when you are speaking or writing as **today**.

I hope you're feeling better today.

3 'the other day'

The other day means 'a few days ago'.

We had lunch together the other day.

4 talking about a particular day

You usually use **on** to talk about a particular day.

We didn't catch any fish on the first day.

5 'these days' and 'nowadays'

You use **these days** or **nowadays** to talk about the present time, especially when things are different now.

I can afford to do what I want these days.
Nowadays most children watch television.

6 'one day'

You use **one day** to say that something will happen sometime in the future.

I'll come back one day, I promise.

In stories, **one day** means an occasion in the past.

One day, he came home and she wasn't there.

dead

Someone who is **dead** is not alive.

They covered the body of the dead woman.
He was shot dead.

You can also say that animals or plants are **dead**.

He knew the spider was dead.
Mary threw away the dead flowers.

> Don't confuse **dead** with **died**. **Died** is the past tense form and past participle of the verb **die**. Don't use **died** as an adjective.
>
> *Sadly, she died of cancer.*

deal

1 'a great deal' and 'a good deal'

A great deal or **a good deal** of something is a lot of it.

He spent a great deal of time thinking about it.
She drank a good deal of coffee that night.

> These expressions can only be used with uncountable nouns. You can talk, for example, about **a great deal of money**, but not about '~~a great deal of apples~~'.

2 'deal with'

When you **deal with** a problem, you give your attention to it.

They learned to deal with any sort of emergency.

The past tense form and past participle of **deal** is *dealt* /delt/, not 'dealed'.

She dealt with the problem quickly.

defeat → see **win – defeat – beat**

definitely → see **surely – definitely – certainly**

delay – cancel – postpone – put off

1 'delay'

If a situation **delays** an event, it causes that event to start at a later time.

Rain delayed the start of the match.

If a plane, train, ship, or bus **is delayed**, it is late leaving or arriving.

The flight has been delayed one hour, due to weather conditions.

2 'cancel', 'postpone' and 'put off'

If you **cancel** an event, you decide that it will not take place.

We cancelled our trip to Washington.
The match was cancelled yesterday because of bad weather.

If you **postpone** or **put off** an event, you decide that it will take place at a later time.

The meeting has been postponed until Tuesday.
We can't put off the decision much longer.

demand

Demand can be a noun or a verb.

1 used as a countable noun

A **demand** for something is a firm request for it.

There were demands for better services.

2 used as an uncountable noun

Demand for a product or service is the amount of it that people want.

Demand for organic food rose by 10% last year.

3 used as a verb

If you **demand** something, you ask for it in a very firm way.

They are demanding higher wages.
I demand to see a doctor.

When **demand** is a verb, don't use 'for' after it. Don't say, for example, '~~They are demanding for higher wages~~.'

deny

1 'deny'

If someone accuses you of something and you **deny** it, you say that it is not true. **Deny** must be followed by an object, a clause beginning with 'that', or an '-ing' form.

He denied that he was involved in the crime.
Rob denied stealing the bike.
Don't deny it!

2 'say no'

If someone answers 'no' to an ordinary question, don't say that they 'deny' what they are asked. Don't say, for example, '~~I asked him if Sue was at home, and he denied it~~'. Say 'I asked him if Sue was at home, and he **said no**'.

I asked her if she wanted to go to the cinema and she said no.

3 'refuse'

If someone says that they will not do something, don't say that they 'deny' it. Say that they **refuse to do** it or **refuse**.

He refused to talk to me.
I asked him to apologise, but he refused.

depend

1 'depend on'

If you **depend on** or **depend upon** someone or something, you need them in order to do something.

Julie seemed to depend on Simon more and more.
The health of the forest depends upon the health of each individual tree.

If one thing **depends on** another thing, it is affected by that thing.

The cooking time depends on the size of the potato.

> **Depend** is never an adjective. Don't say, for example, that someone or something is 'depend' on another person or thing. You say that they are **dependent** on that person or thing.

2 'depending on'

You use **depending on** to say what else affects a situation.

They cost £20 or £25 depending on the size.

3 'it depends'

Sometimes people answer a question by saying '**It depends**'. They usually then explain what else affects a situation.

'What time will you arrive?' – 'It depends. If I go by train, I'll arrive at 5 o'clock. If I go by bus, I'll be a bit later.'

describe

1 used with a noun phrase

When you **describe** someone or something, you say what they are like.

Can you describe your son?

2 used with a clause

You can use **describe** in front of a clause beginning with **what**, **where** or **how**.

The man described what he had seen.
He described how he escaped from prison.

You can say that you **describe** something **to** someone.

She described the feeling to me.

> You must use **to** in sentences like this. Don't say, for example, '~~She described me the feeling~~'.

desert – dessert

1 'desert' – noun

When **desert** is a noun, it is pronounced /dezət/. A **desert** is a large area of land where there is almost no water, trees or plants.

...the Sahara Desert.

2 'desert' – verb

When **desert** is a verb, it is pronounced /dɪzɜːt/. When people or animals **desert** a place, they all leave it.

Poor farmers are deserting their fields and coming here looking for jobs.

If you **desert** someone, you leave them and do not help or support them.

All our friends have deserted us.

3 'dessert'

Dessert /dɪzɜːt/ is sweet food served at the end of a meal.

For dessert there was ice cream.

despite

→ see **in spite of – despite**

dessert

→ see **desert – dessert**

different

If one thing is **different from** another, it is not like the other thing.

London was different from most European capital cities.

Many British people say that one thing is **different to** another. **Different to** means the same as **different from**.

Morgan's law books were different to theirs.

Some people think this is incorrect. In conversation, you can use either **different from** or **different to**, but in writing it is better to use **different from**.

In American English, you can say that one thing is **different than** another.

I am no different than I was 50 years ago.

dinner

→ see topic: **Meals**

disagree – refuse

1 'disagree'

If you **disagree with** someone, you have a different opinion from them.

He disagreed with her.
O'Brien disagreed with the suggestion that his team played badly.

You say that two or more people **disagree about** something.

They always disagree about politics.

2 'refuse'

If someone says that they will not do something, don't say that they 'disagree' to do it. Say that they **refuse** to do it.

He refused to give them any money.

disappear

If someone or something **disappears**, they go where they cannot be seen.

I saw him disappear round the corner.

> Don't use 'disappeared' as an adjective. If you cannot find something because it is not in its usual place, don't say that it 'is disappeared'. Say that it **has disappeared**.
>
> *My keys have disappeared!*

discuss

If you **discuss** something with someone, you talk to them seriously about it.

She could not discuss his school work with him.
We need to discuss what to do.

> **Discuss** is always followed by a noun or a clause. You cannot say, for example, '~~I discussed with him~~' or '~~They discussed~~'.

discussion – argument

1 'discussion'

If you have a **discussion** with someone, you have a serious conversation with them.

After the lecture there was a discussion.

You say that you have a discussion **about** something or a discussion **on** something.

We had long discussions about our future plans.
We're having a discussion on sporting activities.

2 'argument'

Don't use **discussion** to talk about a situation where people get angry with each other. This is usually called an **argument**.

I said no, and we had a big argument over it.

disturb – disturbed

1 'disturb'

If you **disturb** someone, you interrupt what they are doing by talking or making a noise.

If she's asleep, don't disturb her.
Sorry to disturb you, but can I use your telephone?

2 'disturbed'

The adjective **disturbed** has a different meaning. If someone is **disturbed**, they are very upset.

He was disturbed by the news of the attack.

disturbed → see **disturb – disturbed**

do

Do can be an auxiliary verb or a main verb. Its other forms are *does, doing, did, done.*

1 used as an auxiliary verb

When a question or a negative statement is in the present simple or the past simple, and the other verb in the sentence is not 'be', you often use **do**.

Did you enjoy the film?
Where do you come from?
I didn't see you there.

When you want to tell someone not to do something, you use the negative form **don't** followed by another verb.

Don't leave!
Don't forget to lock the door when you leave.

When you want to tell someone to do something, you normally use a verb without 'do', for example 'Come here' or 'Sit down'. However, you can add **do** when you want to show that you strongly want someone to do something, or when you are being very polite.

Do be careful.
Do sit down.

2 used as a main verb

Do is used as a main verb to say that someone performs an action, activity, or task.

I did a lot of work this morning.
After lunch we did the dishes.
The children should do their homework before dinner.

You use **do** when you are asking someone what their job is.

'What do you do?' – 'I'm a teacher.'

> You don't normally use 'do' when you are talking about creating or constructing something. Instead you use **make**.
>
> *Marcella made a delicious cake.*
>
> → see **make**

dozen

You can call twelve things **a dozen** things.

...a dozen eggs.
He found more than a dozen men having dinner.

> You use **a** in front of **dozen**. Don't talk about '~~dozen things~~'.

You can put a number in front of **dozen**. For example, you can talk about 48 things as **four dozen** things.

They wanted three dozen cookies for a party.

> Use the singular form **dozen** after a number. Don't talk about '~~two dozens cups and saucers~~'. Also, don't use 'of' after **dozen**. Don't say '~~two dozen of cups and saucers~~'.

In conversation, you can use **dozens** to talk about a very large number of things. Use **dozens of** in front of a noun.

She borrowed dozens of books.

dream

Dream can be a noun or a verb. The past tense form and past participle of the verb is either *dreamed* /driːmd, dremt/ or *dreamt* /dremt/.

Dreamt is not usually used in American English.

1 used as a noun

A **dream** is a series of events that you see in your mind while you are asleep.

In his dream he was sitting in a theatre watching a play.

You say that someone **has** a dream.

The other night I had a strange dream.

> You don't usually say '~~I dreamed a dream~~'.

A **dream** is also something that you often think about because you would like it to happen.

My dream is to have a house in the country.

2 used as a verb

When someone has a dream while they are asleep, you can say that they **dream** something happens or **dream that** something happens.

I dreamed Marnie was in trouble.
Daniel dreamed that he was back in Minneapolis.

You can also say that someone **dreams about** someone or something or **dreams of** them.

Last night I dreamed about you.
One night I dreamt of him.

When someone would like something to happen very much, you can say that they **dream of having** something or **dream of doing** something.

He dreamed of having a car.
Every small boy dreamed of becoming an engine driver.

> Don't say '~~He dreams to have a car.~~'

dress

1 'get dressed'

When someone **gets dressed**, they put on their clothes.

I got dressed and then had breakfast.

2 'dress up'

If you **dress up**, you put on different clothes so that you look smarter than usual. People **dress up** in order to go, for example, to a wedding or to an interview.

You don't need to dress up for dinner.

If someone **dresses up as** someone else, they wear the kind of clothes that person usually wears.

She dressed up as a princess for the party.

> You only use **dress up** to say that someone puts on clothes that are not their usual clothes. If someone normally wears attractive clothes, don't say that they 'dress up well'. Say that they **dress well**.
>
> *He dresses well and has an expensive car.*

drink

Drink can be a verb or a noun.

1 used as a transitive verb

When you **drink** a liquid, you take it into your mouth and swallow it. The past tense form of **drink** is *drank*, not 'drinked' or 'drunk'.

I drank some of my tea.

The past participle is *drunk*.

He hadn't drunk enough water.

2 used as an intransitive verb

If you use **drink** without an object, you are usually talking about drinking alcohol.

You shouldn't drink and drive.

If you say that someone **drinks**, you mean that they often drink too much alcohol.

Her mother drank, you know.

If you say that someone **does not drink**, you mean that they do not drink alcohol at all.

She said she didn't smoke or drink.

3 used as a countable noun

A **drink** is an amount of liquid that you drink.

I asked her for a drink of water.

Drinks usually means alcoholic drinks.

The drinks were served in the sitting room.

4 used as an uncountable noun

Drink is alcohol.

There was plenty of food and drink at the party.

during

You use **during** or **in** to talk about something that happens from the beginning to the end of a period of time.

We often get storms during the winter.
This music was popular in the 1960s.

→ see **in**

You use **during** to say that something happens while an activity takes place.

I fell asleep during the performance.

You can use **in** in sentences like this, but the meaning is not always the same. For example, 'What did you do **during** the war?' means 'What did you do while the war was taking place?', but 'What did you do **in** the war?' means 'What part did you play in the war?'

You usually use **in** to say when a single event happened.

Mr Tyrie left Hong Kong in June.

> Don't use **during** to say how long something lasts.
> Don't say, for example, '~~I went to Wales during two weeks~~'.
> Say 'I went to Wales **for** two weeks'.

duty

→ see **obligation – duty**

each

→ see topic: **Talking about men and women**

each other – one another

1 'each other'

You use **each other** to show that each member of a group does something to or for the other members. For example, if Simon likes Louise and Louise likes Simon, you say that Simon and Louise like **each other**.

You can use **each other** after a verb or a preposition.

We help each other a lot.
Terry and Mark looked at each other angrily.

2 'each other's'

You can use the possessive form **each other's** before a noun.

They read each other's essays.

3 'one another'

In more formal English, some people use **one another** instead of 'each other'. There is no difference between the two phrases.

They smiled at one another.

earn

→ see **gain – earn**

easily

→ see **easy – easily**

east

→ see topic: **North, South, East and West**

eastern

→ see topic: **North, South, East and West**

easy – easily

1 'easy'

If something is **easy**, it is not difficult. The comparative and superlative forms of **easy** are *easier* and *easiest*.

Losing weight is not easy.
This is much easier than it sounds.

You can say that **it is easy to do** something, or that something **is easy to do**. For example, instead of saying 'Cleaning this room is easy', you can say '**It is easy to clean** this room' or 'This room **is easy to clean**'.

It is easy to use this software.
The museum was easy to find.

2 'easily'

The adverb form of **easy** is **easily**. The comparative and superlative forms of **easily** are *more easily* and *most easily*.

Most students found jobs easily at the end of their course.
The data can be processed more easily with this program.

educate

→ see **bring up – raise – educate**

effect

→ see **affect – effect**

either

1 'either'

When one negative statement follows another, you can put **either** at the end of the second one.

I can't play tennis and I can't play golf either.

2 either ... or

You use **either** and **or** when you you want to say that there are only two possibilities to choose from. You put **either** in front of the first possibility and **or** in front of the second one.

I was expecting you either today or tomorrow.
Either she goes or I go.

electric – electrical

1 'electric'

You use **electric** in front of nouns to talk about particular machines that use electricity.

I switched on the electric fire.

2 'electrical'

You use **electrical** when you are talking generally about machines or systems that use electricity. For example, you talk about **electrical equipment** and **electrical appliances**.

...electrical appliances such as dishwashers and washing machines.

You also use **electrical** to talk about people or organizations that work with electricity.

He is an electrical engineer.

electrical

→ see **electric – electrical**

else

1 with 'someone', 'somewhere' and 'anything'

You use **else** after words such as **someone**, **somewhere**, or **anything** to talk about another person, place, or thing.

If you don't like this, try something else.
'I saw Susan at the park.' – 'Did you see anybody else?'

2 with 'wh'-words

You can use **else** after words such as **when**, **where** and **what**. For example, if you ask '**What else** did you read?', you are asking what other things somebody read, besides the things that they already mentioned.

What else did you get for your birthday?
Who else was there?
Where else did you go last summer?

However, don't use 'else' after 'which'.

embarrassed → see **ashamed – embarrassed**

emigration – immigration – migration

1 'immigrate', 'immigration', 'immigrant'

If you **immigrate** to a country, you go to live in that country permanently.

They immigrated to Israel.

People who immigrate are called **immigrants**.

The company employs several immigrants.

The act of immigrating is called **immigration**.

The government is changing immigration laws.

2 'emigrate', 'emigration', 'emigrant'

If you **emigrate**, you leave your own country and go to live permanently in another country.

His parents emigrated from Canada in 1954.

People who emigrate are called **emigrants**. The act of emigrating is called **emigration**. These words are less frequent than **immigrant** and **immigration**.

3 'migrate', 'migration', 'migrant'

When people **migrate**, they move to another place for a short period of time in order to find work.

Many people migrated to Jakarta to look for work.

People who migrate are called **migrants** or **migrant workers**.

...migrants looking for a place to live.
In South America there are three million migrant workers.

The act of migrating is called **migration.**

end

1 'end'

When something **ends**, it stops. When you **end** something, you cause it to stop.

The meeting ended.
He wanted to end their friendship.

2 'end up'

In conversation, you use **end up** to say what happens to someone at the end of a series of events. You can say that someone **ends up** in a place, that they **end up** with something, or that they **end up** doing something. Don't use **end up** in formal writing.

They ended up back at the house again.
We missed our train, and we ended up taking a taxi.

enjoy

1 enjoy something

If you **enjoy** something, you like it.

I enjoyed the holiday.

Don't say '~~I enjoyed.~~'

2 enjoy yourself

If you have had a pleasant experience, you can say that you **enjoyed yourself**.

I've enjoyed myself very much.

People often say '**Enjoy yourself**' to someone who is going to an occasion such as a party.

Enjoy yourself on Wednesday.

3 enjoy doing something

You can say that someone **enjoys doing** something.

I enjoy going for long walks.

Don't say '~~I enjoy to go for long walks.~~'

enough

1 used in front of a noun

You use **enough** in front of a noun to say that there is as much of something as you need. You can use **enough** in front of countable and uncountable nouns.

There are enough bedrooms for the family.
We don't have enough money.

2 used after adjectives and adverbs

You use **enough** after an adjective or adverb to say that something is acceptable.

Is the soup hot enough for you?
The student isn't trying hard enough.

If you want to say that someone has as much of a quality as they need in order to do something, you add a 'to'-infinitive after **enough**.

She is old enough to work.

3 used as a pronoun

Enough can be used on its own as a pronoun.

They aren't doing enough.

equally

You use **equally** in front of an adjective to say that a person or thing has as much of a quality as someone or something else.

He was an excellent pianist. Irene was equally brilliant.

Don't use 'equally' in front of **as** when you are comparing things. Don't say, for example, '~~He is equally as tall as his brother~~'. Say 'He is **as tall as** his brother'.

He was just as shocked as I was.

→ see **as ... as**

equipment

Equipment consists of all the tools and machines you need for a particular activity.

...kitchen equipment.
...tractors and other farm equipment.

> **Equipment** is an uncountable noun. Don't talk about '~~equipments~~' or '~~an equipment~~'. You can talk about a single item as a **piece of equipment.**
>
> *This radio is an important piece of equipment.*

even

1 showing that something is surprising

You use **even** to show that what you are saying is surprising. You put **even** in front of the surprising part of your statement.

Even Anthony enjoyed it.
Rob still seemed happy, even after the bad news.

You can put **even** in front of nouns, verbs and prepositions. However, **even** goes after an auxiliary verb, not in front of it.

You don't even like him very much.
I couldn't even see the road.

2 used with comparatives

You can use **even** in front of a comparative in order to make it stronger.

Our car is big, but theirs is even bigger.

3 'even if'

You use **even if** to say that a particular fact does not change anything.

Even if you disagree with her, you should listen to her ideas.

4 'even though'

Even though means 'although'.

She wasn't embarrassed, even though she had made a mistake.

> If you begin a sentence with **even if** or **even though**, don't use 'yet' or 'but' as well. Don't say, for example, '~~Even if you disagree with her, yet you should listen to her ideas~~'.

evening

→ see topic: **Times of the day**

eventually – finally

1 'eventually'

When something happens after a lot of delays or problems, you can say that it **eventually** happens.

Eventually they got to the hospital.
I found Victoria Avenue eventually.

> Don't use 'eventually' when you mean that something might be true. Use **possibly** or **perhaps**.
>
> *Perhaps he'll call later.*

2 'finally'

Don't confuse **eventually** with **finally**. You say that something **finally** happens after you have been waiting for it for a long time.

When John finally arrived, he said he'd lost his way.

You can also use **finally** to introduce a final point, ask a final question, or mention a final item.

Combine the flour and the cheese, and finally, add the milk.

ever

1 'ever'

Ever is used in negative sentences, questions, and comparisons. It means 'at any time in the past' or 'at any time in the future'.

I don't think I'll ever trust people again.
Have you ever played football?
I'm happier than I've ever been.

2 'yet'

Don't use 'ever' in questions to ask whether an expected event has happened. Don't say, for example, '~~Has the taxi arrived ever?~~'. The word you use is **yet**.

Don't use 'ever' in negative sentences to say that an expected event has not happened so far. Don't say, for example, '~~The taxi has not arrived ever~~'. The word you use is **yet**.

Have you had your lunch yet?

→ see **yet**

3 'always'

Don't use 'ever' in positive sentences to say that there was never a time when something was not true. Don't say, for example, '~~I've ever been happy here~~'. The word you use is **always**.

I've always been happy here.

→ see **always**

4 'still'

Don't use 'ever' to say that something is continuing to happen. Don't say, for example, '~~When we left Detroit, it was ever raining~~'. The word you use is **still**.

She still lives in London.

→ see **still**

every

1 used for talking about members of a group

You use **every** in front of a singular noun to show that you are talking about all the members of a group.

She spoke to every person at the party.
I agree with every word Peter says.

2 used for saying how often something happens

You also use **every** with expressions of time such as **day** and **afternoon**, in order to show how often something happens.

They met every week.
There is a staff meeting every Monday.

3 'every' and 'all'

You can often use **every** or **all** with the same meaning. For example, '**Every** room has a view of the sea' means the same as '**All** rooms have a view of the sea'.

Every is followed by the singular form of a noun, whereas **all** is followed by the plural form.

Every child needs love.
All children like to play.

Every and **all** do not have the same meaning when they are used with expressions of time. For example, if you do something **every morning**, you do it regularly each morning. If you do something **all morning**, you spend the whole of one morning doing it.

He goes running every day.
I was busy all day.

→ see also topic: **Talking about men and women**

everybody → see topic: **Talking about men and women**

everyday – every day

1 'everyday'

An **everyday** event is ordinary and not unusual. **Everyday** life means ordinary events that happen to people in general.

...the everyday problems of living in the city.
Computers are a central part of everyday life.

2 'every day'

Every day has a different meaning. If something happens **every day**, it happens regularly each day.

Shanti asked the same question every day.

every day

→ see **everyday – every day**

everyone

→ see topic: **Talking about men and women**

everywhere

If you say that something happens **everywhere**, you mean that it happens in all parts of a place.

We searched everywhere.

Don't use 'to' in front of **everywhere**. Don't say, for example, '~~He has been to everywhere~~'. Say 'He has been **everywhere**'.

In informal American English, **every place** is often used instead of 'everywhere'.

Every place we go, people ask us the same questions.

exam – test

An **exam** or a **test** is a series of questions that you answer to show how much you know about a subject. **Exam** is a less formal but very common word for **examination**.

I was told the exam was difficult.
All candidates have to take an English language test.

Exam and **test** are very similar. For important, formal situations it is more common to use **exam**. For more informal situations, for instance in a school class, it is more common to use **test**.

A **test** is also a series of actions that you do to show how well you are able to do something.

She hasn't taken her driving test yet.

You say that people **take**, **sit** or **do** an exam or a test.

Many children want to take these exams.
Students must sit an entrance exam.
We did another test.

> Don't use 'make'. Don't say, for example, '~~We made another test.~~'

If someone is successful in an exam or a test, you say that they **pass** it.

Larry passed his university exams when he was sixteen.
I passed my driving test in Holland.

> If you take an exam or test and you do not know the result, don't say you 'pass' it. To **pass** an exam or a test always means to be successful in it.

If someone is unsuccessful in an exam or a test, you say that they **fail** it.

He failed the written paper.
I think I've failed the test.

example

1 'example'

An **example** is something that shows what other things in a group are like. You say that one thing is an example **of** another thing.

This building is a fine example of traditional architecture.

When someone mentions an example, you say that they are **giving** an example.

Could you give me an example?
Let me give you an example of the sort of thing that happens.

> Don't say '~~Could you say me an example~~?'

2 'for example'

When you mention an example of something, you often say **for example**.

Japan, for example, has two languages.

> Don't say '~~by example~~'.

except

1 used with noun phrases

You use **except** to show that you are not including a particular thing, person, or group in your statement. You usually use **except** in front of a noun or a pronoun.

All the boys except Peter started to laugh.
There's nobody that I really trust, except him.

You can use **except for** in the same way.

The room was empty, except for a television.

2 used with clauses

You can use **except** in front of a clause beginning with **when**, **while**, **where**, **what**, or **that**.

I'm much better now, except that I still have a headache.
I don't know anything about Judith except what her mother told me.

Don't confuse **except** with **besides** or **unless**. **Besides** means 'in addition to'.

What languages do you know besides Arabic and English?

You use **unless** to say what will happen if another thing does not happen.

I won't speak to you unless you apologize.

→ see **unless**

→ see also **accept – except**

excited – exciting

1 'excited'

If someone is very happy about an enjoyable or special event that is going to take place, you say that they are **excited**.

He was so excited he couldn't sleep.
Hundreds of excited children were waiting for us.

You say that someone is **excited about** something, or **excited about doing** something.

I'm very <u>excited about playing</u> football again.

2 'exciting'

Don't confuse **excited** with **exciting**. An **exciting** book or film is full of action, and an **exciting** idea or situation makes you feel very enthusiastic.

The film is <u>exciting</u>, and also very scary.
It isn't a very <u>exciting</u> idea.

exciting

→ see **excited – exciting**

excuse

Excuse can be a noun or a verb. When it is a noun, it is pronounced /ɪkskj<u>uː</u>s/. When it is a verb, it is pronounced /ɪkskj<u>uː</u>z/.

1 used as a noun

An **excuse** is a reason that you give in order to explain why you did something or did not do something.

They are trying to find <u>excuses</u> for their failure.

You say that someone **makes** an excuse.

I <u>made</u> an <u>excuse</u> and left the meeting early.

> Don't say 'I ~~said an excuse~~.'

2 used as a verb

If you **excuse** someone for doing something wrong, you are not angry with them.

Please <u>excuse</u> my bad handwriting.

3 'Excuse me'

You can use '**Excuse me**' as a way of politely apologizing. For example, you can say '**Excuse me**' when you are interrupting someone, when you want to get their attention, or when you want to get past them.

Excuse me, but are you Mr Hess?

4 'apologize'

You use **apologize** to talk about the act of saying sorry. If you want to tell someone that you are unhappy or ashamed about something you have done that has hurt them, you say 'I'm sorry' or 'I apologize'.

She apologized for being so mean to Rudolph.

exist

If something **exists**, it is actually present in the world.

It is clear that a serious problem exists.
They walked through my bedroom as if I didn't exist.

> When **exist** has this meaning, don't use it in the progressive. Don't say, for example, '~~It is clear that a serious problem is existing~~'.

You also use **exist** to say that someone manages to live with very little food or money.

How can we exist out here?

When **exist** has this meaning, it can be used in the progressive.

People were existing on a hundred grams of bread a day.

expect

1 'expect'

If you **expect** that something will happen, or if you **expect** something **to** happen, you believe that it will happen.

He expects to lose his job.
We expect that they will win.

If you **expect** that something is true, you believe that it is probably true.

I expect they've gone.

If someone asks if something is true, you can say '**I expect so**'.

'Will Joe be here at Christmas?' – 'I expect so.'

> Don't say '~~I expect it~~'.

If you **are expecting** someone, you believe that they will arrive soon. If you **are expecting** something, you believe that it will happen soon.

They were expecting Wendy and the children.
We are expecting rain.

> When you use **expect** like this, don't use 'to' after it.

2 'wait for'

Don't confuse **expect** with **wait for**. If you **are waiting for** someone or something, you are staying in the same place and not doing things until they arrive.

He sat on the bench and waited for Jill.

→ see **wait**

expensive

If something is **expensive**, it costs a lot of money.

This is very expensive equipment.
This magazine was more expensive than the others.

> Don't say that the price of something is 'expensive'. Say that it is **high**.
>
> *The price is much too high.*
> *Consumers are paying higher prices for these products.*

experience – experiment

1 'experience'

If you have **experience** of something, you have seen it, done it, or felt it.

Do you have any teaching experience?

An **experience** is something important that happens to you.

Moving house can be a difficult experience.

You say that someone **has** an experience.

I had a strange experience last night.

> Don't say '~~I made a strange experience~~'.

2 'experiment'

Don't use 'experience' to talk about a scientific test that someone does in order to discover or prove something. The word you use is **experiment**.

Laboratory experiments show that Vitamin D may slow cancer growth.

You usually say that someone **conducts** an experiment.

We decided to conduct an experiment.

> Don't say '~~We decided to make an experiment~~'.

experiment → see **experience – experiment**

explain

If you **explain** something, you give details about it so that people can understand it.

He explained the law in simple language.

You say that you explain something **to** someone.

We explained everything to the police.
Let me explain to you about Jackie.

> You must use **to** in sentences like this. Don't say, for example, '~~Let me explain you about Jackie~~'.

You use **explain** followed by **that** to say that someone tells someone else the reason for something.

I explained that I was trying to write a book.

explode – blow up

1 'explode'

When a bomb **explodes**, it bursts loudly and with great force, often causing a lot of damage.

A bomb exploded in the capital yesterday.

You can say that someone **explodes** a bomb.

He exploded the bomb in his bag.

2 'blow up'

If someone destroys a building with a bomb, you say that they **blow** it **up**.

He wanted to blow the place up.

> Don't say '~~He wanted to explode the place.~~'

F

fabric

Fabric is cloth that you use for making things like clothes and bags.

This shirt is made from beautiful soft fabric.

Don't confuse **fabric** and **factory**. A **factory** is a building where machines are used to make things.

Rafael works in a carpet factory.

fact

1 'fact'

A **fact** is a piece of knowledge or information that is true.

The report is full of facts and figures.

> Don't talk about '~~true facts~~'. Don't say, for example, '~~These facts are true~~'.

2 'the fact that'

You can talk about a whole situation with the phrase **the fact that**.

He tried to hide the fact that he had failed.

> You must use **that** in sentences like these. Don't say, for example, '~~He tried to hide the fact he had failed~~'.

3 'in fact'

You use **in fact** if you want to give more information about a statement, especially if the new information is surprising.

I don't watch television; in fact, I no longer own a TV.

fair – fairly

1 'fair'

If you say that something is **fair**, you mean that everyone is treated in the same way.

I wanted everyone to get fair treatment.
It's not fair – she's got more than me!

2 'fairly'

The adverb form of 'fair' is **fairly**.

We solved the problem quickly and fairly.

Fairly also has a completely different meaning. It can mean 'to quite a large degree'.

The information was fairly accurate.
I wrote the first part fairly quickly.

Don't use **fairly** in front of a comparative form. Don't say, for example, '~~The train is fairly quicker than the bus~~'. In conversation, say 'The train is **a bit** quicker than the bus'. In writing, use 'The train is **somewhat** quicker than the bus'.

Golf's a bit more expensive.
The results were somewhat lower than expected.

fairly

→ see **fair – fairly**

fall

When something or someone **falls**, they move quickly towards the ground by accident. The past tense form of **fall** is *fell*. The past participle is *fallen*.

He fell and hurt his leg.
A cup fell on the floor.

When you are talking about people or tall objects, you often use **fall down** or **fall over** instead of 'fall'.

She fell down in the mud.
A tree fell over in the storm.

When rain or snow **falls**, it comes down from the sky.

Rain was beginning to fall.

Don't say that someone 'falls' something. Don't say, for example, '~~She screamed and fell the tray~~'. Say 'She screamed and **dropped** the tray'.

Careful! Don't drop it!

Don't say that someone 'falls' a person. Don't say, for example, '~~He bumped into the old lady and fell her~~'. Say 'He bumped into the old lady and **knocked** her **down**' or 'He bumped into the old lady and **knocked** her **over**'.

I got knocked over by a car when I was six.

→ see also topic: **Seasons**

familiar

1 'familiar'

If someone or something is **familiar**, you recognize them because you have seen or heard them before.

There was something familiar about him.

2 'familiar to'

If something is **familiar to** you, you know it well.

Her name is familiar to millions of people.

3 'familiar with'

If you are **familiar with** something, you know or understand it well.

I am familiar with his work.

far

→ see **away – far**

fault

→ see **blame – fault**

favourite

Your **favourite** thing or person is the one you like more than all the others.

What is your favourite film?
Her favourite writer is Hans Christian Andersen.

> Don't use 'most' with **favourite**. Don't say, for example, '~~What is is your most favourite film?~~'.

The American spelling of 'favourite' is **favorite**.

feel

The past tense form of **feel** is *felt*, not 'feeled'.

1 awareness

If you **can feel** something, you are aware of it through your body.

I can feel a pain in my foot.

Use **can** in sentences like these. Don't say, for example '~~I feel a pain in my foot~~'. Also, don't use a progressive form. Don't say '~~I am feeling a pain in my foot~~'.

2 touching

When you **feel** an object, you touch it in order to find out what it is like.

The doctor felt his pulse.

3 impressions

The way something **feels** is the way it seems to you when you hold it or touch it.

The blanket felt soft.
How does it feel? Warm or cold?

When you use **feel** like this, don't use a progressive form. Don't say, for example, '~~The blanket was feeling soft~~'.

4 emotions and sensations

You can use **feel** with an adjective to talk about experiencing an emotion or physical sensation. When you use **feel** like this, you use either a simple or a progressive form.

I feel lonely.
I'm feeling terrible.
She was feeling hungry.

When you use **feel** in this way, don't use a reflexive pronoun. Don't say, for example, '~~I feel myself lonely~~'.

fetch

→ see **bring – take – fetch**

few – a few

1 'a few'

You use **a few** in front of a plural noun to show that you are talking about a small number of people or things.

I'm having a dinner party for a few close friends.
Here are a few ideas that might help you.

2 'few'

You can also use **few** without 'a' in front of a plural noun, but it has a different meaning. It emphasizes that there is only a small amount of something. For example, if you say 'I have **a few** friends', you mean that you have some friends. However, if you say 'I have **few** friends', you mean that you do not have enough friends and you are lonely.

3 'not many'

In conversation, people do not usually use **few** without 'a'. Instead they use **not many**. For example, instead of saying 'I have few friends', people usually say 'I **haven't got many** friends' or 'I **don't have many** friends'.

They haven't got many books.
I don't have many visitors.

> Don't use 'few' or 'a few' with uncountable nouns. Don't say, for example, '~~Would you like a few more milk in your tea?~~' Say 'Would you like **a little** more milk in your tea?'
>
> → see **little – a little**.

finally

→ see **eventually – finally**

find – find out

1 'find': result of a search

If you **find** something you have been looking for, you see it or learn where it is. The past tense form and past participle of **find** is *found*, not 'finded'.

The police searched the house and found a gun.

If you cannot see the thing you are looking for, you say that you **cannot find** it or that you **can't find** it.

I think I'm lost – I can't find the bridge.

2 'find': noticing something

You can use **find** to say that you notice an object somewhere.

Look what I've found!

3 'find': opinions and feelings

You can use **find** to give your opinion about something. For example, if you think that something is funny, you can say that you **find it funny**.

I find his behaviour extremely rude.
I found it easy.

> You cannot use **find out** for any of these meanings.

4 'find out': obtaining information

You use **find out** to talk about learning the facts about something.

Have you found out who broke the photocopier?
I found out the train times.

find out

→ see **find – find out**

fine – finely

1 'fine' used to mean 'very good'

You can use **fine** to say that something is very good.

From the top there is a fine view of the countryside.

2 'fine' used to mean 'satisfactory'

You can also use **fine** to say that something is satisfactory or acceptable.

'Would you like more milk in your coffee?' – 'No, this is fine.'

If you say that you are **fine**, you mean that your health is satisfactory.

'How are you?' – 'Fine, thanks.'

When you use **fine** to mean 'satisfactory', don't use 'very' in front of it. You can use **just** instead.

Everything is just fine.

In conversation, you can use **fine** as an adverb to mean 'satisfactorily' or 'well'.

We got on fine.

> Don't use 'finely' in sentences like these. Don't say, for example, '~~We got on finely~~'.

3 'fine' and 'finely' used to mean 'very thin'

You can also use **fine** to say that something is very thin, or has very thin parts.

She has very fine hair.

You can use **finely** as an adverb with this meaning.

...finely chopped meat.

finely

→ see **fine – finely**

finish

When something **finishes**, it ends.

The concert finished at midnight.

When you **finish** what you are doing, you reach the end of it.

She finished her dinner and went to bed.

You can say that someone **finishes doing** something.

I finished reading your book last night.

> Don't use a verb in the infinitive in sentences like this. Don't say, for example, '~~I finished to read your book last night.~~'

first – firstly

1 'first' used as an adjective

The **first** thing of a kind is the one that comes before all the others.

January is the first month of the year.

2 'first' used as an adverb

If an event happens before other events, you say that it happens **first**.

Ralph spoke first.

> Don't say '~~Ralph spoke firstly~~'.

3 'first', 'firstly' and 'first of all'

You can use **first**, **firstly** or **first of all** to introduce the first thing that you want to say or the first thing in a list.

First, mix the eggs and flour.
There are two reasons why I'm angry. Firstly you're late, and secondly, you've forgotten your homework.
First of all, I'd like to thank you all for coming.

> Don't say '~~firstly of all~~'.

4 'at first'

When you are contrasting something at the beginning of an event with something that happened later, you say **at first**.

At first I was surprised.
At first I thought that the shop was empty, then I saw a man in the corner.

> Don't use 'firstly' in sentences like these.

firstly

→ see **first – firstly**

first name – forename – given name – Christian name

1 'first name'

Your **first name** is the name that you were given when you were born, that comes before your surname.

Do all your students call you by your first name?

2 'forename'

On official forms, **forename** is sometimes used instead of 'first name'.

3 'given name'

In American English, **given name** is sometimes used instead of 'first name' or 'forename'.

What is your given name?

4 'Christian name'

In British English, some people use **Christian name** instead of 'first name'. However, this can be offensive to people who are not from a Christian family.

fit – suit

1 'fit'

If clothes **fit** you, they are the right size.

The dress fits her well.

In British English, the past tense form of **fit** is *fitted*. In American English, the past tense form is *fit*.

The boots fitted James perfectly.
The pants fit him well and were very comfortable.

2 'suit'

If clothes make you look attractive, don't say that they 'fit' you. Say that they **suit** you.

You look great in that dress, it really suits you.

floor – storey – ground

1 'floor'

The **floor** of a room is the flat part that you walk on.

The book fell to the floor.

A **floor** of a building is all the rooms on a particular level. You say that something is **on** a particular floor.

My office is on the second floor.

In British English, the floor that is level with the ground is called the **ground floor**. The floor above it is called the **first floor**, the floor above that is the **second floor**, and so on.

In American English, the floor that is level with the ground is called the **first floor**, the floor above it is the **second floor**, and so on.

2 'storey'

Storey is also used for a level of a building. It is usually used to talk about how high a building is.

...a house with four storeys.

3 'ground'

Don't call the surface of the earth the 'floor'. Call it the **ground**.

The ground was very wet.

foot

→ see topic: **Transport**

for

→ see **since – for**
→ see topic: **Meals**

forename

→ see **first name – forename – given name – Christian name**

forget

1 'forget'

The past tense form of **forget** is *forgot*, not 'forgetted'. The past participle is *forgotten*.

If you **forget** something such as a key or an umbrella, you do not remember to take it with you when you go somewhere.

Sorry to disturb you – I forgot my key.

> You cannot use the verb 'forget' to say that you have put something somewhere and left it there. Instead you use the verb **leave**.
>
> *I left my bag on the bus.*

If you **have forgotten** something that you knew, you cannot remember it.

I have forgotten his name.

If you **forget** something, or **forget about** something, you stop thinking about it.

He helped me to forget about my problems.

2 'forget to'

If you **forget to do** something, you do not do it because you do not remember it at the right time.

She forgot to lock her door one day and two men got in.
Don't forget to call me.

Don't use an '-ing' form in sentences like these. Don't say, for example, '~~She forgot locking her door.~~'

free – freely

1 no payment

If something is **free**, you can have it or use it without paying for it.

The coffee was free.
...free school meals.

The adverb you use with this meaning is **free**, not 'freely'. For example, say 'Pensioners can travel **free** on the buses'. Don't say '~~Pensioners can travel freely on the buses~~'.

2 not busy

If you are **free** at a particular time, you are not busy. **Free time** is time when you are not busy.

They spend most of their free time reading.
Are you free on Tuesday?

3 no controls

You use 'free' as an adjective to describe activities that are not controlled by rules or other people.

They are free to bring their friends home at any time.

Don't use 'free' as an adverb with this meaning. Use **freely**.

They all express their opinions freely in class.

freely

→ see **free – freely**

friend

1 'friend'

Your **friends** are people you know well and like spending time with. You can call a friend who you know very well a **good friend** or a **close friend**.

He's a good friend of mine.
A close friend told me about it.

If someone has been your friend for a long time, you can call them an **old friend**.

I went to visit an old friend from school.

1 'be friends with'

If someone is your friend, you can say that you are **friends with** them.

You used to be good friends with him, didn't you?
I also became friends with Melanie.

friendly

A **friendly** person is kind and pleasant.

The staff are very friendly and helpful.

If you are **friendly to** someone or **friendly towards** someone, you are kind and pleasant to them.

The women had been friendly to Lyn.

> **Friendly** is never an adverb. Don't say, for example, '~~He behaved friendly~~'. Say 'He behaved **in a friendly way**'.
>
> *She smiled at him in a friendly way.*

Don't confuse **friendly** and **sympathetic**. If you have a problem and someone is **sympathetic**, they show that they care and would like to help you.

When I told her how I felt, she was very sympathetic.

→ see **sympathetic – nice – likeable**

frightened → see **afraid – frightened**

from

1 receiving things

When you are talking about the person who wrote you a letter or sent a message to you, you say that the letter or message is **from** that person.

He received a message from his boss.

2 'come from'

If you **come from** a particular place, you were born there, or it is your home.

I come from Scotland.

3 distance

You can use **from** when you are talking about the distance between places.

How far is the hotel from here?

4 time

If something happens **from** a particular time, it begins to happen at that time.

Breakfast is available from 6 a.m.

Don't use 'from' to say that something started to happen at a particular time in the past and is still happening now. Don't say, for example, '~~I have lived here from 1984~~'. Say 'I have lived here **since** 1984'.

He has been a teacher since 1998.

→ see **since – for**

front

1 'front'

The **front** of a building is the part that faces the street or that has the building's main entrance.

We knocked on the front door.

2 'in front of'

If you are between the front of a building and the street, you say that you are **in front of** the building.

People were waiting in front of the art gallery.

Don't say that you are '~~in the front of~~' a building.

3 'opposite'

If there is a street between you and the front of a building, don't say that you are 'in front of' the building. Say that you are **opposite** it.

The hotel is opposite a railway station.

Speakers of American English usually say **across from** rather than **opposite**.

Stinson has rented a home across from his parents.

frontier

→ see **border – frontier**

fruit

Fruit is usually an uncountable noun. Oranges, bananas, grapes, and apples are all **fruit**.

Fresh fruit and vegetables provide fibre and vitamins.
...fruit imported from Australia.

Don't use 'fruits' to talk about several oranges, bananas, etc. Use **some fruit**. For example, say 'I'm going to the market to buy some **fruit**'. Don't say '~~I'm going to the market to buy some fruits~~'.

...a table with some fruit on it.

full

If something is **full of** things or people, it contains a very large number of them.

...a garden full of pear and apple trees.
His office was full of people.

Don't use any preposition except **of** after **full** in sentences like these.

fun – funny

1 'fun'

If something is **fun**, it is pleasant, enjoyable, and not serious.

The course is interesting and it's also fun.

If you **have fun**, you enjoy yourself.

The children had fun at the party.

> **Fun** is an uncountable noun. Don't say that someone '~~has funs~~' or '~~has a fun~~'.

If you want to say that something is very enjoyable, you can say that it is **great fun** or **a lot of fun**.

The game was great fun.

> Don't use 'big' with **fun**. Don't say, for example, '~~The game was big fun~~'.

2 'funny'

If something is **funny**, it is amusing and makes you smile or laugh.

He told funny stories.

You can also say that something is **funny** if it is strange or surprising.

Have you noticed anything funny about this plane?

funny

→ see **fun – funny**

furniture

Furniture is the large objects in a room, such as tables and chairs.

She arranged the furniture.

> **Furniture** is an uncountable noun. Don't talk about '~~a furniture~~' or '~~furnitures~~'. You can talk about 'a piece of furniture'.
>
> *Each piece of furniture matched the style of the house.*

G

gain – earn

1 'gain'

If you **gain** something, you gradually get it.

This gives you a chance to gain experience.

2 'earn'

Don't say that someone 'gains' money for their work. The word you use is **earn**.

She earns two hundred pounds a week.

generally – mainly

1 'generally'

Generally means 'usually' or 'in most cases'.

Wool and cotton blankets are generally cheapest.

2 'mainly'

Don't use 'generally' to say that something is true about most of something, or about most of the people or things in a group. The word you use is **mainly**.

The African people living here are mainly from Mali.

gently – politely

1 'gently'

If you do something **gently**, you do it carefully and softly, in order to avoid hurting someone or damaging something.

I shook her gently and she opened her eyes.

2 'politely'

Don't use 'gently' to say that someone behaves with good manners. The word you use is **politely**.

He thanked me politely.

get

→ see **arrive – reach – get to**
→ see topic: **Transport**

give

The past tense form of **give** is *gave*. The past participle is *given*.

1 things

If you **give** someone something, you offer it to them and they take it. You can give someone something, or give something **to** someone.

She gave Minnie the keys.
He gave the letter to Mary.

If you use **it** for the thing given, it must go before the person it is given to. Say 'He **gave it to his father**'. Don't say '~~He gave his father it.~~'

2 information

You also say that you **give** someone information, advice, etc., or that you **give** information, advice etc. **to** someone.

The pilot gave us no information about what was happening.
She gives career advice to young people.

3 expressions and gestures

When **give** is used to describe expressions and gestures, the expression or gesture goes in front of the person it is directed at.

He gave her a smile.
As he passed me, he gave me a wink.

> Don't use 'to' in sentences like these. Don't say, for example, '~~He gave a smile to her.~~'

→ see also **offer – give – invite**

given name

→ see **first name – forename – given name – Christian name**

glad – happy – cheerful

1 'glad'

If you are **glad** about something, you are pleased about it. Don't use **glad** in front of a noun. Use it after a linking verb such as **be**, **seem** or **feel**.

I'm so glad that she won the prize.

→ see also topic: **Adjectives that cannot be used in front of nouns**

2 'happy'

You can also say that you are **happy** about something when you are pleased about it.

She was happy that his sister was coming.

Happy can also be used to describe someone who is contented and enjoying life, either most of the time, or on a particular occasion. 'Glad' cannot be used with this meaning.

She always seemed such a happy woman.

3 'cheerful'

If someone shows that they are happy by smiling and laughing a lot, you say that they are **cheerful**.

Our postman is always cheerful and polite.

glasses

A person's **glasses** are two pieces of glass in a frame which they wear to help them to see better. **Glasses** is a plural noun and must be followed by a plural verb.

My glasses are broken.

You can also say **a pair of glasses**.

Gretchen took a pair of glasses off the desk.

go

The past tense form of **go** is *went*. The past participle is *gone*.

1 describing movement

When you talk about moving or travelling somewhere, you often use the verb **go**.

I went to Stockholm.
Celia had gone to school.

→ see also **come**

2 leaving

Go is used to say that someone or something leaves a place.

Our train went at 2.25.

3 talking about activities

You can use **go** with an '-ing' form to talk about activities.

Let's go shopping!

You can also use **go** with **for** and a noun phrase to talk about activities.

He went for a walk.

4 'go and'

To **go and** do something means to move somewhere in order to do it.

I'll go and see him in the morning.

5 'be going to'

You use **be going to** to talk about what someone will do or what will happen in the future.

She told him she was going to leave her job.
The weather is going to get worse.

go on

1 'go on' + '-ing' form

If you **go on doing** something, you continue to do it.

They just ignored me and went on talking.

2 'go on' + 'to'-infinitive

If you **go on to do** something, you do it after doing something else.

He later went on to form a successful computer company.

greet

→ see **salute – greet**

ground

→ see **floor – storey – ground**

grow

The past tense form of **grow** is *grew*. The past participle is *grown*.

1 'grow'

When children or young animals **grow**, they become bigger or taller.

Has he grown any taller?

2 'grow up'

When someone **grows up**, they gradually change from a child into an adult.

He grew up in Cambridge.

gymnasium

A **gymnasium** is a building or large room with equipment for doing physical exercise. In conversation, people usually call it a **gym**.

I go to the gym twice a week.

Don't use **gymnasium** to talk about a school for older pupils. In Britain, this kind of school is called a **secondary school**.

In America, it is called a **high school**.

H

hair

Hair can be a countable or an uncountable noun.

1 used as a countable noun

Each of the fine threads that grow on your head and body is a **hair**. You can talk about several of these things as **hairs**.

...the black hairs on the back of his hands.

2 used as an uncountable noun

You talk about all the hairs on your head as your **hair**, not your 'hairs'.

I washed my hands and combed my hair.

hand

Your **hand** is the part of your body at the end of your arm. It includes your fingers and your thumb.

You talk about a particular person's hand as **his hand**, **her hand**, or **my hand**, not 'the hand'.

The man held a letter in his hand.

However, if you say that someone does something to someone else's hand, you usually use **the**.

Dad took Mum by the hand.

happen

1 'happen'

Something that **happens** takes place without being planned.

Then a strange thing happened.

> **Happen** does not have a passive form. Don't say '~~A strange thing was happened~~'.

2 'take place', 'occur'

You use **happen** after words like 'something', 'thing', 'what', or 'this'. After nouns with a more exact meaning, you usually use **take place** or **occur**.

The discussions took place in Paris.
The crash occurred at night.

Don't say that a planned event 'happens'. Say that it **takes place**.

The first meeting took place on 9 January.

3 'happen to'

When something **happens to** someone or something, it takes place and affects them.

I wonder what's happened to Jeremy?

> In sentences like this, don't use any preposition except **to** after **happen**.

happy

→ see **glad – happy – cheerful**
→ see **lucky – happy**

hard – hardly

1 'hard' as an adjective

If something is **hard**, it is not easy to do.

Looking after three babies is very hard work.

2 'hard' as an adverb

If you work **hard** or try **hard**, you do it with a lot of effort.

Many old people have worked hard all their lives.

3 'hardly'

Hardly is an adverb that has a totally different meaning from **hard**. You use **hardly** to say that something is only just true. For example, if someone **hardly** speaks, they do not speak much. If something is **hardly** surprising, it is not very surprising.

Nick hardly slept that night.

If you use an auxiliary verb with **hardly**, you put it first. You say, for example, 'I **can hardly** see'. Don't say '~~I hardly can see~~'.

We could hardly move.

> Don't use 'not' with **hardly**. Don't say, for example, '~~I did not hardly know him~~'. Say 'I **hardly** knew him'.

4 'hardly ever'

If something **hardly ever** happens, it almost never happens.

I hardly ever spoke to them.

hardly

→ see **hard – hardly**

have

The other forms of **have** are *has, having, had*.

1 'have to'

If someone **has to** do something, they must do it.

I have to speak to your father.

→ see **must**

2 actions and activities

You can use **have** in front of a noun phrase to talk about an action.

Have a look at this!
We had dinner together.

→ see also **take**

> You use a progressive form to say that an activity is taking place. For example, you say 'He **is having** a bath at the moment'. Don't say '~~He has a bath at the moment~~'.
>
> *The children are having a party.*

3 owning things, relationships and appearances

You can use **have** to show that someone owns something.

He had a small hotel.

You also use **have** to talk about friends and family.

Do you have any brothers or sisters?

You use **have** to talk about a person's appearance or character.

You have beautiful eyes.

> Don't use a progressive form when you are talking about owning things, relationships or appearances. For example, don't say '~~He is having a small hotel~~' or '~~You are having beautiful eyes~~.'

→ see also topic: **Meals**

have got

1 how to use 'have got'

You can often use **have got** in spoken English with the same meaning as 'have'. You do not usually pronounce **have got**, **has got**, and **had got** in full. You use **'ve got**, **'s got**, or **'d got** instead.

I've got her address.
He's got a beard now.

2 owning things, relationships, and appearances

Have got is mostly used to talk about owning things, relationships, and appearances.

I've got a rather unusual house.
She's got two sisters.
He's got a lovely smile.

3 illness

You often use **have got** to talk about illnesses.

I've got an awful cold.

4 when you don't use 'have got'

You don't use **have got** for all meanings of **have**.
Don't use it when you are talking about an event or action.
For example, don't say '~~I've got a bath every morning~~'.
Say 'I have a bath every morning'.

Don't use **have got** in formal English.

American speakers do not usually use **have got**. Instead they use **have**.

5 negatives

In negative sentences, **not**, or usually **n't**, goes between **have** and **got**.

I haven't got any more paper.

6 questions

In questions, you put the subject between **have** and **got**.

Have you got enough money for a taxi?

hear

If you **can hear** a sound, you know about it because it has reached your ears.

I can hear a car.

> Use **can** in sentences like this. Say, for example, 'I **can hear** a radio'. Don't say '~~I hear a radio~~'. Also, don't use a progressive form. Don't say '~~I am hearing a radio~~'.

The past tense form and past participle of **hear** is *heard* /hɜːd/. If you want to say that someone was aware of something in the past, you use **heard** or **could hear**.

She heard another sound.
She could hear music in the distance.

help

1 'help' as a transitive verb

If you **help** someone, you make something easier for them. You can use it with an infinitive, with or without 'to'. For example, you can say 'I **helped him to move** the desk' or 'I **helped him move** the desk', which means exactly the same.

He helped us to raise a lot of money.
I helped him fix his car.

2 'help' as an intransitive verb

You can also use **help** without an object, followed by an infinitive with or without 'to'. If someone **helps do** something or **helps to do** it, they help other people to do it.

My mum helps cook the meals for the children.
Dora helped to carry the boxes.

> Don't use an '-ing' form after **help**. Don't say, for example, '~~I helped moving the desk~~' or '~~I helped him moving the desk~~'.

3 'cannot help'

If you **cannot help** doing something, you cannot stop yourself from doing it.

I couldn't help laughing when I saw her face.

> Don't use a 'to'-infinitive after **cannot help**. Don't say, for example, '~~I couldn't help to laugh when I saw her face~~'.

here

1 'here'

You use **here** to talk about the place where you are.

We can come here at any time.

> Don't use 'to' in front of **here**. Don't say, for example, '~~We can come to here at any time~~'.

2 'here is' and 'here are'

You can use **here is** or **here are** at the beginning of a sentence when you want to show or give something to someone. You use **here is** in front of a singular noun and **here are** in front of a plural noun.

Here's your coffee.
Here are the addresses that you need.

high – tall

1 'high'

You use **high** to describe things which measure a long way from the bottom to the top. For example, you talk about a **high hill** or a **high wall**.

...the high mountains of northern Japan.

2 'tall'

You use **tall** to describe things that are high but not very wide. So, for example, you talk about a **tall tree** or a **tall chimney**.

...a field of tall waving grass.

You always use **tall** when you are talking about people.

...a tall handsome man.

3 another meaning of 'high'

High also means 'a long way above the ground'. For example, you talk about a **high window** or a **high shelf**.

...a large room with a high ceiling.

hire – rent – let

1 'hire' and 'rent'

If you pay to use something for a short time, you can say that you

hire it or **rent** it. **Hire** is more common in British English and **rent** is more common in American English.

We hired a car and drove across the island.
He rented a car for the weekend.

If you pay regularly in order to use something for a long period, you say that you **rent** it. You do not usually say that you 'hire' it.

She rents the house with three other women.

2 'let'

If you rent a building or piece of land from someone, you can say that they **let** it to you. The past tense form and past participle of **let** is *let*, not 'letted'.

The cottage was let to an actress from London.

'Let' is more common in British English. In American English, you use **rent**.

The house was rented to a farmer.

holiday – vacation

1 'holiday'

In British English, you talk about the time that you spend away from work or school as the **holiday** or the **holidays**.

I can't wait for the summer holidays.

You talk about time that you spend away from home enjoying yourself as a **holiday**.

I went to Marrakesh for a holiday.

When you spend a period of time like this each year, you talk about your **holidays**.

Where are you going for your holidays?

You usually use a determiner (a word like **a**, **that**, **your** or **my**) in front of **holiday** or **holidays**. Don't say, for example, '~~I went to Marrakesh for holidays~~'.

If you are **on holiday**, you do not have to go to school or work, or you are spending some time away from home enjoying yourself.

Remember to turn off the gas when you go on holiday.

2 'vacation'

The usual American word for a period of time spent away from work or school, or away from home enjoying yourself, is **vacation**.

Harold used to take a vacation at that time.

home

→ see **house – home**

homework – housework

1 'homework'

Homework is work that pupils take home to do in the evening or at the weekend. You say that pupils **do** homework.

He never did any homework.

2 'housework'

Housework is work that you do to keep a house clean and tidy.

He does most of the housework.

Use **do**, not 'make', with both **homework** and **housework**. Don't say, for example, '~~He never made any homework~~.' Both **homework** and **housework** are uncountable nouns. Don't talk about ~~a homework~~ or ~~houseworks~~.

hope

1 basic meaning

If you **hope** that something is true or will happen, you want it to be true or to happen.

He hoped she wasn't going to cry.
I sat down, hoping they wouldn't notice me.

2 'I hope'

You often use **I hope** to show that you want something to be good or successful. After **hope** you can use either a future form or the present simple. For example, you can say '**I hope you'll enjoy** the film' or '**I hope you enjoy** the film'.

I hope you'll enjoy your stay in Britain.
I hope you get well soon.

3 'I hope so'

If someone asks you whether something is true, or whether something will happen, you can answer 'yes' by saying '**I hope so**', or 'no' by saying '**I hope not**'.

'Will you be home at six?' – 'I hope so.'
'Have you lost the ticket?' – 'I hope not.'

> Don't say '~~I hope it~~' or '~~I don't hope so~~'.

hospital

→ see topic: **Places**

house – home

1 'house'

You use **house** to talk about the building where you live.

We bought this house because of the garden.

> You do not usually say '~~I am going to my house~~' or '~~She was in her house~~'. You say 'I am going **home**' or 'She was **at home**'.
>
> *Brody arrived home a little before five.*
> *I'll finish the report at home.*

2 'home'

Your **home** is the place where you live or feel that you belong. **Home** can be used to talk about a person's house or to a town, a region, or a country.

My father works away from home.
Dublin will always be home to me.

> Don't use 'to' directly in front of **home**. Don't say, for example, '~~We went to home~~'. Say 'We went **home**'.

housework → see **homework – housework**

how – what

1 'how': ways of doing things

You use **how** when you are talking about the way something is done.

How do you spell his name?
This is how I make a vegetable curry.

2 'how': asking about someone's health

You use **how** with **be** to ask about someone's health.

How are you?
How is she? All right?

3 'how': asking about opinions

You use **how** with **be** to ask someone whether something was good.

How was your trip?
How was the smoked fish?

4 'what': asking for a description

Don't use 'how' to ask for a description of something or someone. For example, if you want a description of someone's boss, don't say '~~How is your boss?~~' Say '**What** is your boss **like?**'

What's his mother like?
What is Fiji like?

however

1 'however'

You use **however** when you want to add something new and different or unexpected after what you've already said.

Losing at games doesn't matter to some women. Most men, however, can't stand it.

You also use **however** to say that it is not important how something is done.

You can do it however you want.

2 'how ever'

Sometimes people use **ever** after **how** at the beginning of a question. They do this to show that they are surprised. For example, instead of saying 'How did you get here?', they say

'**How ever** did you get here?'

How ever did you find me?

Write **how ever** as two separate words. Don't write, for example, '~~However did you find me?~~'

how much

You use **how much** when you are asking about the price of something. For example, you say '**How much** is that T-shirt?'

I like that dress – how much was it?

Don't use 'how much' and 'price' in the same sentence. Don't say, for example, '~~How much is the price of that T-shirt?~~'

Only use **how much** with **be** when you are asking about the price of something. Don't use it to ask about other amounts of money. Don't say, for example, '~~How much is his income?~~' Say '**What is his income?**', '**What does he earn?**', or '**How much does he earn?**'

Similarly, don't say '~~How much is the temperature outside?~~' or '~~How much is the population of Tokyo?~~' Say '**What** is the temperature outside?' or '**What** is the population of Tokyo?'

What is the speed limit here?

hundred – thousand – million

A hundred or **one hundred** is the number 100.

A thousand or **one thousand** is the number 1,000.

A million or **one million** is the number 1,000,000.

You usually say that there are **a hundred/thousand/million** things.

We'll give you a thousand dollars for the story.

You say **one hundred/thousand/million** things when you want to emphasize the number, or when you want to be very clear and precise.

Over one thousand students applied.
The total amount was one hundred and forty-nine pounds and thirty pence.

Don't add '-s' to **hundred**, **thousand** or **million** when you put another number in front of them.

There are more than two hundred languages spoken in Nigeria.

hurt

→ see topic: **Adjectives that cannot be used in front of nouns**

I

I

You use **I** to talk about yourself. **I** is the subject of a verb. You always write it with a capital letter.

I like your dress.

You can also use **I** as part of the subject of a verb. For example, you can say '**My friend and I** are going to Sicily'. Always mention the other person first. Don't say '~~I and my friend are going to Sicily~~'.

My brothers and I are musicians.

if

1 possible situations

You use **if** to talk about a possible situation.

You can go if you want to.

You use **if** to talk about something that might happen in the future. You use a verb in the present simple.

He might win – if he's lucky.

> Don't use a future form in sentences like this. Don't say, for example, '~~He might win – if he will be lucky~~.'

You use **if** to talk about things that sometimes happened in the past. You use a past form of the verb.

They sat outside if it was sunny.

2 unlikely situations

You use **if** to talk about things that will probably not happen. You use the past simple.

If I frightened them, they might run off and I would never see them again.

3 in reported questions

You use **if** when you are reporting a question where the answer is 'yes' or 'no'. For example, if you say to someone 'Can I help you?', you can report this by saying 'I asked her if I could help her.'

He asked me if I spoke French.

ill – sick

1 'ill' and 'sick'

Ill and **sick** are both used to say that someone has a disease or some other problem with their health.

Davis is ill.
Your uncle is very sick.

> Don't say that someone becomes '~~iller~~' or '~~more ill~~'. Say that they become **worse**.
>
> *Each day she felt a little worse.*

2 'be sick'

To **be sick** means to bring up food from your stomach.

I think I'm going to be sick.

3 'feel sick'

To **feel sick** means to feel that you want to be sick.

Flying always makes me feel sick.

imagine

If you **imagine** a situation, you think about it and form a picture or idea of it in your mind.

Try to imagine you're sitting on a cloud.

You can use an '-ing' form after **imagine**.

She could not imagine living without Daniel.

> Don't use a 'to'-infinitive after **imagine**. Don't say, for example, '~~She could not imagine to live without Daniel~~'.

immediately

If something happens **immediately**, it happens without delay.

I have to go to Brighton immediately. It's very urgent.

If something happens **immediately after** something else, it happens as soon as the other thing is finished.

He had to see a customer immediately after lunch.

If something is **immediately above** something else, it is above it and very close to it. You can use **immediately** in a similar way with other prepositions such as **under**, **opposite** and **behind**.

This man sat down immediately behind me.

immigration → see **emigration – immigration – migration**

important

If something is **important**, you feel that you must have, do, or think about it.

This is the most important part of the job.

Don't use 'important' to say that an amount or quantity is very large. Don't talk, for example, about '~~an important sum of money~~'. Use a word such as **considerable** or **significant**.

A considerable amount of rain had fallen.

in

1 used to say where something is

You use **in** to say where someone or something is, or where something happens.

Colin was in the bath.
I wanted to play in the garden.
Mark now lives in Singapore.

→ see also topic: **Places**

2 used to say where something goes

You use **in** to mean 'into a place'.

She opened her bag and put her diary in.

In can sometimes mean 'into'.

She threw both letters in the bin.

→ see **into**

3 used to talk about time

You use **in** to say how long something takes.

He learned to drive in six months.

You use **in** to talk about a particular year, month, season or part of the day.

I was born in 1972.

→ see topic: **Times of the day**
→ see topic: **Seasons**

4 used to mean 'wearing'

You can use **in** to say what someone is wearing.

Who is the woman in the red dress?

→ see **wear – in**

> Don't use 'in' when you are talking about someone's ability to speak a foreign language. Don't say, for example, '~~She speaks in Russian~~'. Say 'She speaks Russian'.
>
> → see **speak – talk**

indoor

→ see **indoors – indoor – outdoors – outdoor**

indoors – indoor – outdoors – outdoor

1 'indoors' and 'outdoors'

Indoors and **outdoors** are adverbs. If something happens **indoors**, it happens inside a building. If it happens **outdoors**, it does not happen in a building.

I spent all the evenings indoors.
School classes were held outdoors.

If you go **indoors**, you go into a building.

Let's go indoors.

> When someone goes out of a building, don't say that they go 'outdoors'. Say that they go **outside**.
> *When they went outside, it was raining.*

2 'indoor' and 'outdoor'

Indoor and **outdoor** are adjectives used in front of a noun. You use **indoor** to describe something that is done or used inside a building and **outdoor** to describe something that is done or used outside.

...indoor swimming pools.
...an outdoor play area.

information – news

1 'information'

Information means facts about someone or something. You say that you **give** people information.

Pat did not give her any information about Sarah.

> Don't use 'tell'. Don't say, for example, '~~Pat did not tell her any information about Sarah.~~'

You refer to information **about** something or **on** something.

I'd like some information about trains, please.
I'm afraid that I have no information on that.

> **Information** is an uncountable noun. Don't talk about '~~an information~~' or '~~informations~~'. You can talk about a **piece of information**.
>
> *I wondered how to use this piece of information.*

2 'news'

Don't use **information** to talk about descriptions of recent events in newspapers or on television or radio. The word you use is **news**.

It was on the news at 8.30.

in spite of – despite

1 'in spite of'

You use **in spite of** when you are talking about a fact that makes the rest of what you are saying sound surprising. The spelling is **in spite of**, not '~~inspite of~~'.

In spite of his illness, my father was always cheerful.

Don't use 'in spite of' as a conjunction. Don't say, for example, '~~In spite of we protested, they took him away~~'. Say '**Although** we protested, they took him away'.

Although he was late, he stopped to buy a sandwich.

2 'despite'

Despite means the same as **in spite of**. Don't use 'of' after **despite**.

Despite their different ages, they were close friends.

instead – instead of

1 'instead'

Instead is an adverb. You use it when you are saying that someone does something rather than doing something else.

Judy did not answer. Instead she looked out of the taxi window.

2 'instead of'

Instead of is a preposition. You use it to introduce something that is in the place of something else.

Why not have your meal at seven o'clock instead of five o'clock?

You can say that someone does something **instead of doing** something else.

You should walk to work instead of driving.

Don't use a 'to'-infinitive in sentences like this. Don't say, for example, '~~You should walk to work instead of to drive~~.'

instead of → see **instead – instead of**

intention → see **meaning – intention – opinion**

interested – interesting

1 'interested'

If you want to know more about something or someone, you can say that you are **interested in** them.

I am very interested in politics.

Don't use any preposition except **in** after **interested**.

If you want to do something, you can say that you are **interested in doing** it.

I was interested in seeing different kinds of film.

Don't use a 'to'-infinitive in sentences like this. Don't say, for example, '~~I was interested to see different kinds of film~~.'

2 'interesting'

Don't confuse **interested** with **interesting**. You say that someone or something is **interesting** because you want to know more about them.

I've met some very interesting people.

interesting

→ see **interested – interesting**

into

You use the preposition **into** to talk about movement of some kind. You use **into** to say where someone or something goes, or where you put something.

I went into the church.

However, in front of **here** and **there**, you use **in**, not 'into'.

Come in here.

After verbs meaning 'put', 'throw', 'drop', or 'fall', you can use **into** or **in** with the same meaning.

William put the letter into his pocket.
He locked the bag and put the key in his pocket.

invite

If you **invite** someone to a party or a meal, you ask them to come to it.

He invited Alexander to dinner.
I invited her to my party.

> You must use **to** in sentences like these. Don't say '~~I invited her my party~~.'

When you ask someone to do something enjoyable, you can say that you **invite** them **to do** it.

He invited Axel to come with him.

> Don't say '~~He invited Axel for coming with him.~~'

→ see also **offer – give – invite**

irritated

→ see **nervous – anxious – irritated**

it

1 used to talk about things

You use **it** to talk about an object, animal, or other thing that has just been mentioned.

...a tray with glasses on it.

2 used to talk about situations

You can also use **it** to talk about a situation, fact, or experience.

I like it here.
She was frightened, but tried not to show it.

> When you use a verb such as **like** or **prefer** with an '-ing' form or a 'to'-infinitive, don't use 'it' as well. For example, don't say '~~I like it, walking in the park~~'. Say 'I like walking in the park'.
>
> *I like being in your house.*
> *I want to be a doctor.*

3 used with verbs like 'be' and 'become'

You use **it** followed by **be** to say what the time, day, or date is.

It's seven o'clock.
It's Sunday morning.

You use **it** followed by a linking verb like **be** or **become** to describe the weather or the light.

It was very cold.
It became dark.

Don't use 'it' followed by 'be' to say that something exists. Don't say, for example, '~~It's a lot of traffic on this road tonight~~'. You say '**There's** a lot of traffic on this road tonight'.

There was no more room in the house.

→ see **there**

it's

→ see **its – it's**

its – it's

1 'its'

You use **its** to show that something belongs to a thing, place or animal.

He held the knife by its handle.
The horse raised its head.

2 'it's'

It's is a short form of 'it is' or 'it has'.

It's three o'clock.
It's been very nice talking to you.

J

jam

→ see **marmalade – jam – jelly**

jelly

→ see **marmalade – jam – jelly**

job

→ see **position – post – job**

journey – trip – voyage

1 'journey'

A **journey** is an occasion when you travel from one place to another.

...a journey of over 2,000 miles.

2 'trip'

A **trip** is an occasion when you travel from one place to another, stay there for a short time, and come back again.

...a business trip to Milan.

3 'voyage'

A **voyage** is a long journey from one place to another in a ship or spacecraft.

The ship's voyage is over.
...the voyage to the moon in 1972.

4 verbs used with 'journey', 'trip' and 'voyage'

You can **make** a journey, trip or voyage, or **go on** a journey, trip or voyage.

He went on a journey to London.
I made a special trip to Yorkshire to visit them.

> Don't use 'do' with any of these words. Don't say, for example, '~~He did a journey to London~~'.

just

You use **just** to say that something happened a very short time ago. British speakers usually use the present perfect with **just**. For example, they say 'I**'ve just** arrived'.

I've just bought a new house.

American speakers usually use the past simple. Instead of saying 'I've just arrived', they say 'I **just** arrived'.

I just broke the pink bowl.

K

keep

The past tense form and past participle of **keep** is *kept*, not 'keeped'.

1 storing

If you **keep** something somewhere, you store it in that place.

Where do you keep your keys?

2 staying in a particular state

You can use **keep** followed by an adjective to talk about staying in a particular state. For example, if you 'keep someone warm', you make them stay warm. If someone 'keeps warm', they stay warm.

The noise outside kept them awake.
They have to hunt for food to keep alive.

3 used with an '-ing' form

Keep can be used in two different ways with an '-ing' form.

You can use it to say that something happens again and again.

The phone keeps ringing.
My mother keeps asking questions.

You can also use it to say that something continues to happen and does not stop.

I turned back after a while, but he kept walking.

know

1 being aware of facts

If you **know** that something is true, you are aware that it is a fact. The past tense form of **know** is *knew*, not 'knowed'. The past participle is *known*.

I knew that she was studying at law school.

> Don't use a progressive form with **know**. Don't say, for example, '~~I am knowing that this is true~~'. Say 'I **know** that this is true'.

2 'I know'

If someone tells you something that you already know, don't say '~~I know it~~'. Say '**I know**'.

'That's not their fault, Peter.' – 'Yes, I know.'

3 being familiar with things and people

If you **know** a person, place, or thing, you are familiar with them.

Do you know David?
He knew London well.
Do you know the poem 'Kubla Khan'?

4 'know how to'

If you **know how to** do something, you have learnt how to do it.

Do you know how to drive?

> You must use **how** in sentences like this. Don't say, for example, '~~Do you know to drive?~~'.

L

lady
→ see **woman – lady**

large
→ see **big – large**

last – lastly

Last can be an adjective or an adverb.

1 'last' used as an adjective

The **last** thing, event, or person is the one that comes after all the others.

He missed the last bus.
They met for the last time just before the war.

2 'last' used as an adverb

If something **last** happened on a particular occasion, it has not happened since then.

We last saw him nine years ago.

If an event is the final one in a series, you can say that it happens **last**. You put **last** at the end of the sentence.

He added the milk last.

3 'last' with time expressions

You use **last** in front of a word such as **week**, **month**, **Christmas** or **autumn** to talk about a date or a period of time before the present one.

I saw him last week.
She died last autumn.

→ see topic: **Times of the day**

> Don't use 'the' before 'last' in this meaning. Don't say, for example, '~~I saw him the last week~~'.

4 'lastly'

Lastly is used for the final item in a list.

Lastly I would like to ask about your future plans.

lastly

→ see **last – lastly**

late – lately

1 'late'

Late can be an adjective or an adverb.

If you are **late** for something, you arrive after the time that was arranged.

I was ten minutes late for my appointment.

You can also say that someone arrives **late**.

Etta arrived late.

> Don't use 'lately' for this meaning of the adverb.

2 'lately'

You use **lately** to say that something has been happening since a short time ago.

We haven't been getting on so well lately.

lately

→ see **late – lately**

lay – lie

1 'lay'

Lay is a transitive verb, and it is also a past tense form of **lie**.

If you **lay** something somewhere, you put it there carefully. The other forms of **lay** are *lays, laying, laid.*

Lay a sheet of newspaper on the floor.
I carefully laid Marianne down on the sofa.

2 'lie'

Lie is an intransitive verb.

If you **lie** somewhere, you are in a flat position, not standing or sitting. The other forms of **lie** in this meaning are *lies, lying, lay, lain.*

She lay on the bed, reading.
The baby was lying on the table.

If you **lie**, you say or write something that you know is not true. When **lie** is used in this meaning, its other forms are *lies, lying, lied.*

He lied to me.
She was sure that Thomas was lying.

learn – teach

1 'learn'

When you **learn** something, you obtain knowledge or a skill as a result of studying or training.

The past tense form and past participle of **learn** is *learned*. In British English, *learnt* is also used.

We first learned to ski in the Alps.
He had never learnt to read and write.

2 'teach'

Don't say that you 'learn' someone something or 'learn' them how to do something. The word you use is **teach**. The past tense form and past participle of **teach** is *taught*, not 'teached'.

Mother taught me how to read.

If you **teach** a subject, you explain it to people as your job.

I taught history for many years.

You can either say that you **teach** someone something or that you **teach** something **to** someone.

...the man that taught us English at school.
I found a job teaching English to a group of adults in Paris.

If you **teach** someone **to do** something, you give them instructions so that they know how to do it.

Boylan taught him to drive.

lend

→ see **borrow – lend**

let

→ see **allow – let**
→ see **hire – rent – let**

let's – let us

1 'let's': making a suggestion

Let's is short for 'let us'. It is used to make suggestions for you and someone else, and is followed by an infinitive without 'to'.

Let's go outside.

If you are saying that you and someone else should not do something, you say **let's not**.

Let's not talk about that.

2 'let us': talking about permission or asking for information

When you are talking about you and someone else being allowed to do something, you use **let us**.

They wouldn't let us sleep.

Let us is also used in the phrase **let us know** to ask for information about something.

Let us know what progress has been made.

let us

→ see **let's – let us**

library – bookshop

1 'library'

A **library** is a building where books are kept for people to use or borrow.

He often went to the public library.

2 'bookshop'

In Britain, a shop where you buy books is called a **bookshop**, not a 'library'. In America, it is called a **bookstore**.

You work in a bookshop, don't you?

lie

→ see **lay – lie**

like

1 'like'

If you **like** someone or something, you find them pleasant or attractive.

She's a nice girl. I like her.
He liked the room, which was large.

> Don't use **like** in progressive forms. Don't say, for example, '~~I am liking her~~'.

If you enjoy an activity, you can say that you **like doing** it.

I like reading.

You can add **very much** to emphasize how much you like someone or something.

I like him very much.
I like driving very much.

> You must put **very much** after the person or thing that you like. Don't say, for example, '~~I like very much driving~~'.

If someone asks you if you like something, you can say 'Yes, I **do**.' Don't say '~~Yes, I like.~~'

'Do you like walking?' – 'Yes I do, I love it.'

> Use **like it** in front of a clause beginning with **when** or **if**. Don't say '~~I like when I can go home early~~'. Say 'I **like it** when I can go home early'.

2 'would like'

You say '**Would you like...?**' when you are offering something to someone, or inviting someone to do something.

Would you like some coffee?
Would you like to meet him?

> Don't say '~~Do you like some coffee?~~'

You can say '**I'd like...**' when you are asking for something in a shop or a café.

I'd like some apples, please.

likeable

→ see **sympathetic – nice – likeable**

listen to

If you **listen to** something or someone, you pay attention to their sound or voice.

I was listening to the radio.
Listen carefully to what he says.

> Use **to** in sentences like these. Don't say, for example, '~~I was listening the radio~~'.

Don't confuse **listen** and **hear**. If you **hear** something, you become aware of it without trying. If you **listen to** something, you deliberately pay attention to it.

I heard a noise.

→ see **hear**

little – a little

1 'a little'

A little is used in front of uncountable nouns to talk about a small quantity or amount of something.

I had made a little progress.

2 'little'

If you use **little** in front of a noun, you are emphasizing that there is not enough of something. For example, if you say 'We got **a little** help from them', you mean that they gave you some help. If you say 'We got **little** help from them', you mean that they did not give you enough help.

It is clear that little progress was made.

3 'not much'

A little and **little** are slightly formal. In conversation, **not much** is used instead. For example, instead of saying 'I have little money', you say 'I **haven't got much** money' or 'I **don't have much** money'.

I haven't got much work to do.
We don't have much time.

lonely

→ see **alone – lonely**

long

1 used to talk about length

You use **long** when you are talking about the length of something.

...an area up to 3000 feet long and 900 feet wide.
How long is that side of the triangle?

2 talking about distance

You use **a long way** to talk about a large distance from one place to another.

It's a long way from here to Birmingham.

> Don't say '~~It's long from here to Birmingham~~'.

In questions or negative sentences, you use **far**.

Is the school far from here?
It was not far to walk back to our hotel.

3 used to talk about time

You use **a long time** to talk about a large amount of time.

We may be here a long time.

> Don't say '~~We may be here long.~~'

In questions or negative sentences, you can use **long** as an adverb to mean 'a long time'.

Are you staying long?

You can also say **too long** or **long enough**.

He's been here too long.
You've been here long enough to know what we're like.

look

1 'look at'

If someone directs their eyes towards something, you say that they **look at** it.

Lang looked at his watch.

When **look** has this meaning, it must be followed by **at**. Don't say, for example, '~~Lang looked his watch~~'.

Don't confuse **look** with **see** or **watch**.

→ see **see – look at – watch**

If you want to say that someone shows a particular feeling when they look at someone or something, you show this using an adverb.

Jenna looked sadly at the floor.

2 used to mean 'seem'

Look can also be used with an adjective to mean 'seem' or 'appear'.

You look very pale.
Seth looked disappointed.

look after – look for

1 'look after'

If you **look after** someone or something, you take care of them.

She will look after the children during their holidays.

2 'look for'

If you **look for** someone or something, you try to find them.

He looked for his shoes under the bed.

→ see also **search – look for**

look for

→ see **look after – look for**
→ see also **search – look for**

look forward to

1 used with a noun

If you **are looking forward to** something that you are going to experience, you are pleased or excited about it.

They were looking forward to the summer holidays.

> Use **to** in sentences like these. Don't say, for example, '~~They were looking forward the summer holidays~~'.

2 used with an '-ing' form

You can say that you **look forward to doing** something.

I look forward to seeing you in Washington.

> Don't use an infinitive after **look forward to**. Don't say, for example, '~~I look forward to see you in Washington.~~'

lot

1 'a lot of' and 'lots of'

You use **a lot of** or **lots of** in front of a noun when you are talking about a large number or amount of people or things. **Lots of** is used in conversation.

A lot of people thought it was funny.
You've got lots of time.

When you use **a lot of** or **lots of** in front of a plural noun, you use a plural form of a verb with it. If you use them in front of an uncountable noun, you use a singular form of the verb.

A lot of people come to our classes.
Lots of time was spent playing with these toys.

2 'a lot'

You use **a lot** without a noun to talk about a large quantity or amount of something.

I've learnt a lot.

You also use **a lot** as an adverb to mean 'to a great extent' or 'often'.

You like Ralph a lot, don't you?
They talk a lot about politics.

lucky – happy

1 'lucky'

You say that someone is **lucky** when something nice happens to them, or when they always seem to have good luck.

The lucky winners were given £5000 each.

2 'happy'

Don't use 'lucky' to say that someone feels pleased and satisfied. The word you use is **happy**.

Sarah's such a happy person – she's always laughing.
Barbara felt very happy.

lunch

→ see topic: **Meals**

M

mad

1 stupid

In conversation, you can say that a stupid action or suggestion is **mad**.

You're going to swim in that water? You must be mad!
That's a mad idea.

2 angry

In conversation, **mad** can also mean 'angry'. If you **go mad**, you become angry. If you are **mad at** someone, you are angry with them. When you use **mad** in this way, don't put it in front of a noun.

Mum went mad when I told her.
He's mad at me because I broke his computer.

> Don't use 'mad' in formal writing.

3 mentally ill

If someone has an illness that makes them behave in strange ways, you should not say that they are 'mad'. You should say that they are **mentally ill**.

Susan is mentally ill.
...the treatment of mentally ill patients.

made from – made of – made out of

1 'made from'

If one thing is **made from** another thing, the first thing is produced from the second thing, and the second thing is changed completely in the process

Most wine is made from grapes.

2 'made of'

If something was used to produce another thing, and it was not completely changed, use **made of**. Don't use 'made from'.

The hut was made of logs.

3 'made out of'

If something was produced from another thing in an unusual way, use **made out of**.

He was wearing a hat made out of an old coat.

made of

→ see **made from – made of – made out of**

made out of

→ see **made from – made of – made out of**

magazine – shop

1 'magazine'

A **magazine** is a thin book with stories and pictures that you can buy every week or every month.

I often read fashion magazines.

2 'shop'

Don't use 'magazine' to talk about a place where you buy things. The word you use is **shop**.

I work in a clothes shop.

mainly

→ see **generally – mainly**

make

The past tense form and past participle of **make** is *made*.

1 doing and saying things

You can use **make** when you want to say that someone does or says something. For example, if someone suggests something, you can say that they **make** a suggestion. If someone promises something, you can say that they **make** a promise.

I made the wrong decision.
In 1978 he made his first visit to Australia.

Here is a list of common nouns that you can use with **make** in this way:

arrangement
choice
comment
decision
effort
mistake
noise
plan
point
promise
sound
speech
suggestion
visit

You use 'make' only when you are mentioning a particular action. When you are talking generally about action, you use **do**.

What have you done?
You've done a lot to help us.

2 creating and producing things

If you **make** an object or substance, you create or produce it.

Sheila makes all her own clothes.

You can also say that someone **makes** a meal or a drink.

I made his breakfast.

→ see **cook**
→ see topic: **Meals**

If you create or produce something for another person, you can say that you **make** someone something, or **make** something **for** someone.

> *I have made you a drink.*
> *My grandmother made this dress for me.*

→ see also **brand – make**

man

→ see topic: **Talking about men and women**

manage – arrange

1 'manage'

If you **manage to do** something, you succeed in doing it.

> *Manuelito managed to escape.*
> *How did you manage to do that?*

Don't use an '-ing' form after **manage.** Don't say, for example, '~~How did you manage doing that?~~'

2 'arrange'

Don't use 'that' after 'manage'. Don't say, for example, that you '~~manage that something is done~~'. Say that you **arrange for something to be done**, or that you **arrange for someone to do something**.

> *He arranged for the parcel to be sent to America.*
> *I arranged for a mechanic to fix the car.*

mankind

→ see topic: **Talking about men and women**

many

1 questions

You use **many** in front of a plural noun when you are asking about numbers of people of things.

How many brothers and sisters do you have?

2 negative statements

You use **not** ... **many** in negative statements when you are talking about a small number of people or things.

He doesn't have many friends.

3 positive statements

You can also use **many** in positive statements when you are talking about a large number of people or things.

Many people disagreed with him.

In conversation, people often use **a lot of** or **lots of** instead of **many**.

I have a lot of books.

4 'many of'

You use **many of** in front of a plural pronoun, or in front of a determiner such as **the** or **his** followed by a plural noun.

Many of them had to leave.
How many of his books have you read?

Don't use 'many' with uncountable nouns. Use **much**.

→ see **much**

marmalade – jam – jelly

1 'marmalade'

Marmalade is a sweet food made from oranges, lemons, limes, or grapefruit. People often spread it on bread.

We had toast and marmalade for breakfast.

2 'jam'

Jam is a sweet food made from other fruit such as blackberries, strawberries, or apricots.

My wife made this delicious strawberry jam.

3 'jelly'

In American English, a food like this is often called **jelly**.

...a raspberry jelly sandwich.

marriage – wedding

1 'marriage'

Marriage is the state of being married, or the relationship between a husband and wife.

He had three children from his first marriage.
They had a happy marriage.

2 'wedding'

You don't usually use 'marriage' to talk about the ceremony in which two people get married. The word you use is **wedding**.

I was invited to Paul and Sue's wedding.

married – marry

1 'married to'

If you are **married to** someone, that person is your husband or wife.

Her daughter is married to a Frenchman.

2 'marry'

When you **marry** someone, you become their husband or wife during a special ceremony.

I want to marry him.

Don't say '~~I want to marry to him~~.'

3 'get married'

You don't usually use 'marry' without a following noun. Don't say, for example, '~~She married~~' or '~~They married~~'. Use **get married**.

I'm getting married next month.
They got married in October 1994.

marry

→ see **married – marry**

match

If one thing has the same colour or pattern as another thing, you say that the first thing **matches** the other thing.

The lampshades matched the curtains.
Do these shoes match my dress?

Don't use 'matches to.' For example, don't say '~~Do these shoes match to my dress?~~'

matter

1 'What's the matter?'

You can say **What's the matter?** to ask about a problem or difficulty.

What's the matter? You seem unhappy.

> Don't use 'the matter' with this meaning in other types of sentence. Don't say, for example, '~~The matter is that we don't know where she is~~'. Say **the problem** or **the trouble**.
>
> *The problem is that she can't cook.*

2 'It doesn't matter'

When someone apologizes to you, you can say '**It doesn't matter**.' Don't say '~~No matter~~'.

'I'm sorry, I've spilled some milk.' – 'It doesn't matter.'

may

→ see **might – may**

me – myself

You use **me** to talk about yourself. **Me** can be the object of a verb or a preposition.

He told me about it.
He looked at me angrily.

> Don't use 'me' when the person who is speaking is both the subject and the object of the verb. Don't say, for example, '~~I got me a drink~~'. Say 'I got **myself** a drink'.
>
> *I made myself some breakfast.*

mean

The past tense form and past participle of **mean** is *meant* /ment/, not 'meaned'.

You use **mean** when you are talking about the meaning of a word or expression.

What does 'software' mean?
'Unable' means 'not able'.

> You must use **does** in sentences like these. Don't say, for example, '~~What means 'software'?~~'

If you **mean to do** something, you intend to do it.

I'm sorry, I didn't mean to hurt you.

> Don't use 'mean' when you are talking about people's opinions or beliefs. Use **think** or **believe**. Don't say, for example, '~~Most of the directors mean he should resign~~'. Say 'Most of the directors **think** he should resign'.
>
> *I think it will rain tomorrow.*
> *Scientists believe that life began four billion years ago.*

meaning – intention – opinion

1 'meaning'

The **meaning** of a word or expression is the thing or idea that it represents.

The word 'set' has many different meanings.

2 'intention'

Don't use **meaning** to say what someone intends to do. Don't say, for example, '~~Her meaning was to leave before midnight~~'. Say 'Her **intention** was to leave before midnight'.

My intention is to retire next year.

3 'opinion'

Don't use 'meaning' to say what someone thinks about something. Don't say, for example, '~~I think he should resign. What's your meaning?~~' Say 'I think he should resign. What's your **opinion**?'

If you want my opinion, I think this is a terrible idea.

media

You can call television, radio, and newspapers **the media**.

He told his story to the media.

In conversation, some people use a singular form of a verb with **the media**.

The media is full of pictures of worried families.

In formal writing, however, you should use a plural form of a verb.

The media have not commented on the story.

memory → see **souvenir – memory**

mention → see **comment – mention – remark**

might – may

You can use **might** or **may** to say that it is possible that something is true or will happen in the future. **May** is more formal than **might**.

They still hope that he might be alive.

It may rain tomorrow.
I might go to London next year.

You can use **could** in a similar way.

'Where's Jack?' – 'He could be upstairs.'

You use **might not** or **may not** to say that it is possible that something is not true. In conversation, you can also use the short form **mightn't**.

He might not be in England at all.
That mightn't be true.

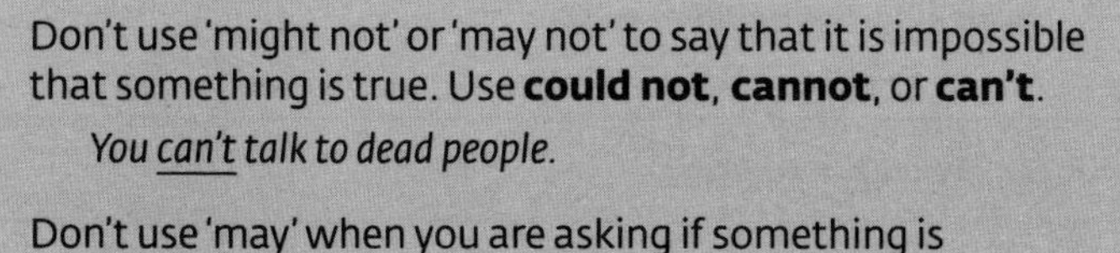

Don't use 'might not' or 'may not' to say that it is impossible that something is true. Use **could not**, **cannot**, or **can't**.

You can't talk to dead people.

Don't use 'may' when you are asking if something is possible. Use **might** or **could**. Don't say, for example, '~~May he be right?~~' Say '**Might** he be right?' or **Could** he be right?'

Could this be true?

→ see **can – could – be able to**

migration → see **emigration – immigration – migration**

million → see **hundred – thousand – million**

mind

1 'make up your mind'

When someone decides to do something, you can say that they **make up their mind** to do it.

She made up her mind to write to Frances.

2 'don't mind'

If you have no objection to doing something, you can say that you **don't mind doing** it.

I don't mind walking.

> Don't say '~~I don't mind to walk~~'.

You can show that you do not object to a situation or suggestion by saying '**I don't mind**'.

'Do you want me to invite Marcin?' – 'I don't mind, if you want to.'
It was raining, but he didn't mind.

> Don't say '~~I don't mind it.~~'

mistake

1 'mistake'

A **mistake** is something that is not correct. You say that someone **makes** a mistake.

He made a terrible mistake.
Tony made three spelling mistakes in his essay.

> Don't use 'do' with **mistake**. Don't say, for example, '~~He did a terrible mistake.~~'

2 'by mistake'

Don't say that someone does something '~~in mistake~~'. Say that they do it **by mistake**.

I called the wrong number by mistake.

3 'fault'

Don't use 'mistake' to talk about something wrong in a machine. Use **fault**.

There's a fault in one of the appliances.

money

Money is the coins or bank notes that you use to buy things.

Cars cost a lot of money.
She spends too much money on clothes.

After **money** you use a singular form of a verb.

Money isn't important.

> **Money** is an uncountable noun. Don't talk about '~~moneys~~' or '~~a money~~'.

more

1 'more' and 'more of'

You use **more** or **more of** to show that you are talking about a larger number of people or things, or an additional number of people or things.

You can use **more** in front of both plural and uncountable nouns.

We sold more cars this year than last year.
We need more information.

You use **more of** in front of a pronoun or a determiner (such as **the** or **his**).

More of them are getting jobs.

He wants to spend more of his time with his children.

2 **'more than'**

If you want to show that there is a larger amount of something than a particular number, you use **more than** in front of the number.

My husband died more than twenty years ago.
Police arrested more than seventy people.

You use a plural form of a verb after **more than**.

More than 100 people were at the party.

3 **'more' used in comparatives**

You can use **more** to form comparative adjectives and adverbs. You use **more** in front of adjectives that have two or more syllables. You use **more** in front of most adverbs that end in '-ly'.

Love is more important than money.
Next time, I will choose more carefully.

> Don't use **more** with adjectives that have only one syllable. Don't say, for example, '~~more tall~~'. Instead, add '-er' to the end of the adjective.
>
> *Krish is taller than me.*

morning → see topic: **Times of the day**

most

1 **'most' and 'most of'**

You use **most** or **most of** to show that you are talking about the largest part or amount of people or things.

You use **most** in front of plural nouns.

Most people think he is a great actor.

You use **most of** in front of a pronoun or a determiner (such as **the** or **his**).

Most of us have strong views on politics.
He spends most of his time in the library.
Most of Roberta's friends came to the party.

> When you use **most** like this, don't use 'the' in front of it. Don't say, for example, '~~She ate the most part of the cake~~'. Say 'She ate **most of** the cake'.

2 'most' used in superlatives

You can also use **most** to form superlative adjectives and adverbs. You use **the most** in front of adjectives that have two or more syllables. You use **most** in front of most adverbs that end in '-ly'.

He is the most intelligent man I know.
The disease spreads most easily in dirty conditions.

> Don't use **the most** with adjectives that have only one syllable. Don't say, for example, '~~the most large room~~'. Instead, use 'the' followed by the adjective with '-est' at the end.
>
> *This is the largest room in the house.*

move

→ see **remove – move**

much

1 'very much'

You use **very much** to emphasize something.

I enjoyed it very much.

> You do not usually use **very much** directly after a verb. Don't say, for example, '~~I enjoyed very much the party~~'. Say 'I enjoyed the party **very much**'.
>
> In positive sentences, don't use **much** without **very**. Don't say, for example, '~~I enjoyed it much~~'. In negative sentences, you can use **much** without 'very'.
>
> *I didn't like him much.*

2 'much' meaning 'often'

You can use **much** in negative sentences and questions to mean 'often'.

She doesn't talk about them much.
Does he come here much?

> Don't use 'much' in positive sentences to mean 'often'. Don't say, for example, '~~He comes here much~~'.

3 'much' used with comparatives

You can use **much** in front of comparative adjectives and adverbs when you want to emphasize the difference between two things.

She was much older than me.
Now I feel much more confident.

4 'much' used in front of a noun

You use **much** in front of an uncountable noun when you are talking about a large amount of something. **Much** is usually used like this in negative sentences, in questions, or after **too**, **so**, or **as**.

There isn't much danger.
Is this going to make much difference?
It gave the President too much power.

In positive sentences, you don't usually use 'much' in this way. Instead, you use **a lot of**.

I did a lot of work at the weekend.

must

1 saying that something is necessary

You can use **must**, **have to** or **need to** in order to say that something is necessary.

I must leave soon.
We have to get up early tomorrow.
I need to make a phone call.

After **must** you use an infinitive without 'to'. Don't say, for example, '~~I must to leave.~~'

If you want to say that something was necessary in the past, you use **had to**. Don't use 'must'.

She had to go to work immediately.

You use **must not** or **mustn't** to say that it is important that something is not done.

We mustn't forget to call Mum.

2 saying that something is not necessary or important

If you want to say that it is not necessary that something is done, use **don't have to**, or **don't need to**. Don't use 'must not' or 'mustn't'.

I don't have to go to work tomorrow.
You don't need to tell me if you don't want to.

3 saying that you believe something is true

You use **must** to say that you strongly believe that something is true, because of particular facts.

Claire's car isn't there, so she must be at work.

> If you want to say that you believe something is not true, you use **cannot** or **can't.** Don't use 'must not' or 'mustn't.'
>
> *The two messages cannot both be true.*
>
> → see **can – could – be able to**

myself

→ see **me – myself**

N

nation

You use **nation** to talk about a country or the people who live there in a rather formal way, especially when you are talking about politics and history.

For a long time, Britain was the most powerful nation on earth.
He appealed to the nation for calm.

In more general situations, you say **country**.

In my job, I travel all over the country.

nationality

You use **nationality** to say what country someone legally belongs to.

He's got British nationality.

nature

1 'nature'

Nature is used to talk about all the animals, plants and things that happen in the world that are not made or caused by people.

The most amazing thing about nature is its variety.

When **nature** has this meaning, don't use 'the' in front of it.

2 'the country'

Use **the country** or **the countryside** to talk about land that is away from towns and cities. Don't use 'nature' for this meaning.

We live in the country.
I've always loved the English countryside.

need

1 'need'

If you **need** something, it is necessary for you to have it.

I need money for food.

> Don't use **need** in progressive forms. Don't say, for example, '~~I am needing money for food.~~'

The negative form is **do not need** or **don't need**.

The letter did not need her signature.
I don't need any help, thank you.

2 'need to do'

If you **need to do** something, it is necessary for you to do it.

You need to work hard if you want to pass your exams.

> You must use **to** in sentences like these. Don't say, for example, '~~You need work hard if you want to pass your exams~~'.

If something is not necessary, you say that you **don't need to do** it or **do not need to do** it.

You don't need to shout.
She does not need to worry about us.

3 'must not'

Must not has a different meaning. If you want to say that it is necessary for someone **not** to do something, you use **must not** or **mustn't**.

You must not accept it.
We mustn't forget the keys.

→ see **must**

neither

You use **neither** or **neither of** to make a negative statement about two people or things.

1 'neither'

You use **neither** in front of a singular noun.

Neither man spoke or moved.

2 'neither of'

You use **neither of** in front of a plural pronoun or a plural noun phrase beginning a determiner such as **the**, **these**, **his** or **its**.

Neither of them spoke for several moments.
Neither of her parents said anything.

3 adding a clause

After a negative statement, you can use **neither** at the beginning of the next sentence or clause to show that this statement is also true for another person or thing.

'I didn't invite them.' – 'Neither did I.'
He'll never forget it, and neither will we.

nervous – anxious – irritated

1 'nervous'

If you are **nervous**, you are frightened or worried about something that you are going to do.

My son is nervous about starting school.

2 'anxious'

If you are worried about something that might happen to someone else, you say that you are **anxious**.

It's time to be going home – your mother will be anxious.

3 'irritated'

If something annoys you, you say that you are **irritated** by it.

Perhaps they were irritated by the sound of crying.

never

→ see topic: **Negatives**

news

→ see **information – news**

next

You use **next** in front of words such as **week**, **month**, **weekend**, **Monday** or **June** to talk about a date or a period of time that is directly after the present one.

I'm getting married next month.
He said he would be seventy-five next April.
Let's have lunch together next Wednesday.

Don't use 'the' in front of **next**. Don't say, for example, '~~I'm getting married the next month.~~'

Don't say that something will happen 'next day'. Say that it will happen **tomorrow**. If you want to say the time of day, you use **tomorrow morning**, **tomorrow afternoon**, **tomorrow evening**, or **tomorrow night**.

Can we meet tomorrow at five?
I'm going down there tomorrow morning.

nice → see **sympathetic – nice – likeable**

night → see topic: **Times of the day**

noise → see **sound – noise**

none → see topic: **Negatives**

no-one → see topic: **Negatives**

north → see topic: **North**, **South**, **East and West**

northern → see topic: **North**, **South**, **East and West**

not → see topic: **Negatives**

nothing → see topic: **Negatives**

now

1 'now'

Now is usually used to contrast a situation in the present with an earlier situation.

She gradually built up energy and is now back to normal.
Now he felt safe.

2 'right now' and 'just now'

In conversation, you use **right now** or **just now** to say that a situation exists at present, although it may change in the future.

I'm very busy just now.

You also use **now** or **right now** to emphasize that something is happening at this moment, or must happen immediately.

She's here with us right now.
He wants you to come and see him now, in his room.

If you say that something happened **just now**, you mean that it happened a very short time ago.

Did you feel the ship move just now?

nowhere

→ see topic: **Negatives**

O

obligation – duty

1 'obligation' and 'duty'

If you say that someone has an **obligation to do** something or a **duty to do** something, you mean that they should do it.

He felt no obligation to invite her for dinner.
Perhaps it was his duty to inform the police of what he had seen.

2 'duties'

Your **duties** are the things that you do as part of your job.

He was busy with his official duties.

occasion – opportunity – chance

1 'occasion'

An **occasion** is a time when something happens.

I remember the occasion vividly.

You often say that something happens **on** a particular occasion.

I met him only on one occasion.

An **occasion** is also an important event, ceremony, or celebration.

The wedding was a happy occasion.

2 'opportunity' and 'chance'

Don't use 'occasion' to talk about a situation in which it is possible for someone to do something. Use **opportunity** or **chance**.

You may have the opportunity of meeting him one day.
She put the phone down before I had a chance to reply.

→ see **chance**

occur

→ see **happen – take place – occur**

of

Of is used to show possession and relationships between people or things.

...the home of a sociology professor.
...the sister of the Duke of Urbino.

You can use **of** in front of **mine**, **his**, **hers**, **ours**, **yours** or **theirs**.

He's a very good friend of mine.
I talked to a colleague of yours recently.

> Don't use 'of' in front of 'me', 'you', 'he', 'she', 'it', 'us' or 'they'. Don't say, for example, '~~the sister of me~~'. Say **my sister**.
>
> *My sister stayed with us last week.*
> *He had his hands in his pockets.*

You don't usually use 'of' in front of names or very short noun phrases. Instead you use **'s**. For example, don't say '~~the car of the man~~'. Say 'the man's car'.

I heard Ralph's voice behind me.

offer – give – invite

1 'offer'

If you **offer** something to someone, you ask them if they want to have it or use it.

He offered me a biscuit.

2 'give'

If you put something in someone's hand and they take it, don't say that you 'offer' it to them. Say that you **give** it to them.

She gave Minnie the keys.

3 'offer to'

If you **offer to do** something, you say that you are willing to do it.

He offered to take her home in a taxi.

4 'invite'

If someone asks you to do something enjoyable, don't say that they 'offer' you to do it. Say that they **invite** you to do it.

He invited me to come to the next meeting.

old

1 talking about age

Old is used to state the age of a person or thing. For example, you say that someone 'is forty years **old**'.

Mary is twenty-nine years old.
The bones were 6,000 years old.

You can also describe someone as, for example, 'a **forty-year-old** man'. Don't say '~~a forty-years-old man~~'.

Sue lives with her five-year-old son.

2 asking about age

You use **how old** to ask about the age of a person or thing.

How old are you?
How old is the Taj Mahal?

3 'old' and 'elderly'

You can also use **old** to describe people or things that have lived or existed for a very long time.

He looked really old.
Her wardrobes were full of old clothes.

Elderly is a more polite word to describe old people.

I look after my elderly mother.

4 old friends

An **old friend** is someone who has been your friend for a long time. He or she is not necessarily an old person.

We visited some old friends.

on

1 used to say where something is

If something is **on** a surface, it is touching or supported by it.

There were several photographs on his desk.

2 used to say where something goes

You can use **on** to say where someone or something falls or is put.

He fell on the floor.
I put a hand on his shoulder.

Onto is used in a similar way.

He threw the envelope onto his desk.

3 used to talk about vehicles

You use **on** with names of vehicles such as a bus, train or aeroplane.

George got on the bus with us.
I met him on the train to Vienna.

4 used to talk about time

You say that something happens **on** a particular day or date.

We are going see the play on Friday.
Caro was born on April 10th.

→ see also topic: **Times of the day**

once

1 'once'

You use **once** to say that something happened one time in the past.

I once spent a night in that hotel.
I have never forgotten her, though I saw her only once.

You also use **once** to say that something was true in the past, although it is no longer true. In this meaning, **once** usually goes after **be** or an auxiliary verb, or at the end of a sentence.

They were once very good friends, but now they never see each other.
She had been a teacher once.

2 'at once'

If you do something **at once**, you do it immediately.

She stopped playing at once.

one

You can use **one** instead of a singular noun when you have already mentioned the noun. For example, instead of saying '~~If you want a drink, I'll get you a drink~~', you say 'If you want a drink, I'll get you **one**'.

She is not a model, but she looks like one.

You can use **one** or **ones** instead of a noun that follows an adjective.

I got this trumpet for thirty pounds. It's quite a good one.
He earns money by buying old houses and building new ones.
Which dress do you prefer? I like the red one.

You can also use **one** after words such as **this**, **each**, **that**, **my** or **another**.

We need a smaller fridge. This one's too big.
She had a plate of soup, then went back for another one.

→ see also topic: **Talking about men and women**

one another → see **each other – one another**

only

You use **only** in front of a noun or **one** to say that something is true about one person, thing, or group and not true about anyone or anything else. In front of **only** you put **the** or a word such as **my**, **his** or **their**.

Grace was the only survivor.
I was the only one smoking.
He scored his only goal in that game.

If you use another adjective or a number before the noun, you put **only** in front of it.

<u>The only English city</u> he enjoyed working in was Manchester.

Only is also an adverb.

→ see topic: **Where you put adverbs**

open

1 used as a verb

If you **open** something such as a door, you move it so that it no longer covers a hole or gap.

He <u>opened</u> the window and looked out.

> When you say that a person **opens** something, you must say what they open. Don't say, for example, '~~I went to the door and opened~~'. Say 'I went to the door and **opened it**'.

2 used as an adjective

When a door or window is not covering the hole or gap it is there to cover, you say that it is **open**.

The door was <u>open</u>.
He was sitting by the <u>open</u> window of the office.

> **Opened** is the past tense form or past participle of the verb **open**. You only use it when you are describing the action of opening something.
>
> *A tall man <u>opened</u> the front door.*
>
> Don't say that a door 'is opened'. Say that it 'is open'.

opinion

→ see **meaning – intention – opinion**
→ see **point of view – view – opinion**

opportunity → see **occasion – opportunity – chance**

opposite

1 used as a preposition

If one building or room is **opposite** another, they are separated from each other by a street or corridor.

The hotel is opposite a railway station.
The bathroom was opposite my room.

If two people are **opposite** each other, they are facing each other, for example when they are sitting at the same table.

Lynn was sitting opposite him.

Speakers of American English usually say **across from** rather than 'opposite' in the above senses.

Stinson has rented a home across from his parents.

2 used as a noun

If two things or people are completely different from each other, you can say that one is **the opposite of** the other.

The opposite of right is wrong.
He was the exact opposite of Herbert.

ought to → see **should – ought to**

outdoor → see **indoor – indoors – outdoor – outdoors**

outdoors → see **indoor – indoors – outdoor – outdoors**

over

1 position

If one thing is **over** another thing, it is directly above it.

There is a mirror over the fireplace.

2 movement

If you go **over** something, you cross it and get to the other side.

James stepped over the dog.

3 age

If someone is **over** a particular age, they are older than that age.

She was well over fifty.

4 time

If something happens **over** a period of time, it happens during that time.

He had flu over Christmas.

→ see also **above – over**

overseas

1 used as an adverb

If you go **overseas**, you visit a foreign country that is separated from your own country by sea.

Roughly 4 million Americans travel overseas each year.

2 used as an adjective

Overseas is used in front of nouns. It has a similar meaning to 'foreign', but is more formal and is used especially when talking about business and politics.

He met the president on a recent overseas visit.

> If you say that someone **is overseas**, you do not mean that they are foreign. You mean that they are visiting a foreign country.

own

1 'own'

You use **own** after a word like **my**, **its** or **our** to emphasize that something belongs to or is connected with a particular person or thing.

I took no notice till I heard my own name mentioned.
Now the industry's own experts support these claims.

You also use **own** to say that something belongs only to the person or thing mentioned.

She says we cannot have our own key to the apartment.
Each room had a style of its own.

2 'on your own'

If you are **on your own**, you are alone.

She lived on her own.

If you do something **on your own**, you do it without any help from anyone else.

We can't solve this problem on our own.

P

pair – couple

1 'a pair of'

A pair of things are two things of the same size and shape that are used together, such as shoes. You can use a singular or a plural verb with this meaning.

A pair of new gloves were lying on the table.
A pair of shoes was on display in the window.

You also use **a pair of** to talk about something with two main parts of the same size and shape, such as trousers, glasses, or scissors. You use a singular form of a verb with this meaning.

She put on a pair of glasses.
Lying on the bed was an old pair of trousers.

2 'a couple of'

You can say that two people or things are **a couple of** people or things. This is slightly informal. You use a plural form of a verb with **a couple of**.

There are a couple of police officers outside.
On the table were a couple of newspapers.

3 'couple'

You say that two people are a **couple** when they are married or are in a romantic relationship. You usually use a plural form of a verb with **couple**.

The couple have two children.

paper

Paper is the material that you write things on or wrap things in.

The students will all be given pencils and paper.

In this meaning, **paper** is an uncountable noun. For one piece, say a **piece of paper**, or if it is a whole piece you can also say a **sheet of paper**.

He wrote his name at the top of a blank sheet of paper.
Rob picked up the piece of paper and gave it to her.

Newspapers are often called **papers**.

I read about his death in the papers.
The Daily News is the country's largest daily paper.

pardon

1 to apologize

You can apologize to someone by saying '**I beg your pardon**'.

Oh, I beg your pardon – I didn't realise you were sitting there.

2 when you have not heard

In British English, you can say **Pardon?** when you did not hear what someone said and you want them to say it again.

'How old is she?' – 'Pardon?' – 'I said how old is she?'

pass

→ see **spend – pass**

past

1 the past

The **past** is the time before the present.

In the past, most babies with the disease died.

2 telling the time

In British English, when you are telling the time, you use **past** to say how many minutes it is after a particular hour.

It's ten past eleven.

American speakers usually say **after**.

It's ten after eleven.

3 going near something

Past is also used as a preposition or adverb to say that someone goes from one side of something to the other.

He walked past Lock's hat shop.
People ran past laughing.

4 'passed'

Don't use 'past' as the past tense form or past participle of the verb **pass**. The word you use is **passed**.

As she passed the library door, the telephone began to ring.

pay

The past tense form and past participle of **pay** is *paid*, not 'payed'.

If you **pay for** something, you give someone money for something you are buying.

Willie paid for the drinks.

You must use **for** after **pay** in sentences like these. Don't say '~~Willie paid the drinks~~'.

If you pay for a meal or a drink for someone else, don't say that you 'pay' them the meal or the drink. Say that you **buy** them the meal or the drink.

I'll buy you lunch.

people – person

1 'people'

People is a plural noun. You use a plural form of a verb after it.

People is used to talk about a group of men, women and children.

There were 120 people at the lecture.
Hundreds of people were killed in the fire.

2 'person'

Person is a countable noun. A **person** is a man, woman, or child.

There was far too much meat for one person.

The usual plural of 'person' is **people**. **Persons** is used only in formal or official situations.

This is an additional payment for persons responsible for a child.

person

→ see **people – person**

persuade

→ see **convince – persuade**

phone

When you **phone** someone, you dial their phone number and speak to them by phone.

I went back to the hotel to phone Jenny.

You can also **phone** a place.

He phoned the police station.

Don't use 'to' after **phone**. Don't say, for example, '~~He phoned to the police station.~~'

place

1 used in descriptions

A **place** is a particular building, room, town, or area.

Richmond is a beautiful place.
Keep your dog on a lead in public places.

2 'there'

Don't use 'that place' to talk about somewhere that has just been mentioned. Don't say, for example, '~~I drove my car into a field and left it in that place~~'. You say 'I drove my car into a field and left it **there**'.

I decided to try Newmarket. I soon found a job there.

3 'room'

Don't use 'place' to talk about space for someone or something to fit into. You use **room** instead.

There's not enough room in the bathroom for both of us.

play

1 children's games

When children **play**, they spend time using their toys or taking part in games.

The kids played on the swings.

2 sports and games

If you **play** a sport or game, you take part in it regularly.

Ray and I play tennis three times a week.
Do you play chess?

3 CDs and DVDs

If you **play** something like a CD, DVD or video, you use a piece of equipment to listen to it or watch it.

She played her CDs too loudly.

4 musical instruments

If you **play** a musical instrument, you produce music from it, or you are able to produce music from it.

Nina was playing the piano when I arrived.
Can you play the guitar?

point of view – view – opinion

1 'point of view'

When you are thinking about one part of a situation, you can say that you are thinking about it from a particular **point of view**.

From a medical point of view he did everything wrong.

A person's **point of view** is the way they feel about something that affects them.

We understand your point of view.
I tried to see things from Frank's point of view.

2 'view' and 'opinion'

Don't call what someone thinks or believes about a particular matter their 'point of view'. Use **view** or **opinion**.

The police's view is that it was an accident.
If you want my honest opinion, I don't think it will work.

You talk about someone's opinions or views **on** or **about** a matter.

He always asked for her opinions on his work.
She has strong views about politics.

You can use phrases such as **in my opinion** or **in his view** to show that something is someone's belief, and is not necessarily a fact.

He's not being very helpful, in my opinion.
In his view, this proposal would be unsuccessful.

police

The police are the official organization responsible for making sure that people obey the law.

He called the police.

Police is a plural noun. You use a plural form of a verb after it.

The police were called to the scene of the crime.

A single member of the police is called a **police officer**, a **policeman**, or a **policewoman**.

He has been a police officer for six years.

politely

→ see **gently – politely**

position – post – job

1 'position' and 'post'

In formal English, someone's regular job is called their **position** or **post**. When a job is advertised, it is often described as a **position** or **post**.

She is well qualified for the post.

2 'job'

In conversation, you use **job**.

He's afraid of losing his job.

possible – possibly

1 'possible'

Possible is an adjective. If something is **possible**, it can be done.

If it is possible to find out where your brother is, we will.

Possible is often used in expressions such as **as soon as possible** and **as much as possible**.

Please make your decision as soon as possible.
I like to know as much as possible about my patients.

You also use **possible** to say that something may be true or correct.

It is possible that he said these things.

2 'possibly'

Possibly is an adverb. You use **possibly** to show that you are not sure about something.

We are looking for a new office, possibly in California.

You also use **could you possibly** to ask someone to do something in a very polite way.

Could you possibly check if Mr Dayton has arrived?

possibly

→ see **possible – possibly**

post

→ see **position – post – job**

postpone

→ see **delay – cancel – postpone – put off**

price – cost

1 'price' and 'cost'

The **price** or **cost** of something is the amount of money you must pay to buy it.

...the price of sugar.
The cost of petrol has gone up.

> Don't use any preposition except **of** after **price** or **cost** in sentences like these.

2 'costs'

The plural noun **costs** is used to talk about the total amount of money that you need to do something such as run a business.

We need to cut costs.

3 'cost' used as a verb

You use **cost** to talk about the amount of money that you must pay for something.

The dress costs $200.

To say how much someone paid for something, you use **cost** followed by the name of the person, followed by the amount of money.

The holiday cost me $800.

The past tense form and past participle of **cost** is *cost*, not 'costed'.

prison

→ see topic: **Places**

problem

1 a difficult situation

A **problem** is a difficult situation.

The distance to work is a problem for me.
They have financial problems.

You can say that you **have problems doing** something.

They are having problems sending emails.

> Don't say that someone 'has problems to do' something. Don't say, for example '~~They are having problems to send emails.~~'

2 'reason'

> Don't use 'problem' with **why** when you are explaining the reason something happened. Don't say, for example, '~~The problem why he couldn't come is that he is ill~~'. Say 'The **reason** why he couldn't come is that he is ill'.
>
> *That is the reason why I find her books boring.*

professor – teacher

1 'professor'

In a British university, a **professor** is the most senior teacher in a department.

He was Professor of English at Strathclyde University.

In an American or Canadian university or college, a **professor** is a senior teacher, but not necessarily the most senior teacher in a department.

2 'teacher'

Don't use 'professor' to talk about a person who teaches at a school. The word you use is **teacher**.

I'm a qualified French teacher.

proper

1 used to mean 'real'

You use **proper** in front of a noun to show that someone or something really is something.

Have you been to a proper doctor?

2 used to mean 'correct'

You also use **proper** in front of a noun to say that something is correct or suitable.

What's the proper word for those things?

Don't use 'proper' to mean that something belongs to someone. The word you use is **own**.

I want to have my own business.

prove – test

1 'prove'

If you **prove** that something is true, you give evidence that shows that it is true.

He was able to prove that he was an American.

2 'test'

When you do something to find out how good or bad someone or something is, don't say that you 'prove' the person or thing. Say that you **test** them.

I will test you on your knowledge of French.
The drug was tested on mice.

put off

→ see **delay – cancel – postpone – put off**

put up with

→ see **bear – can't stand – put up with**

Q

quiet – quite

Quiet is an adjective. Someone or something that is **quiet** makes only a small amount of noise.

Bal spoke in a quiet voice.

A **quiet** place does not have much activity or trouble.

It's a quiet little village.

Don't confuse **quiet** /kwaiət/ with **quite** /kwait/.
→ see **quite**

quite

You use **quite** in front of an adjective or adverb. It means 'fairly, but not very'. For example, if something is 'quite big', it is big, but it is not very big.

He was quite young.

You can also use **quite** in front of **a**, followed by an adjective and a noun.

She was quite a pretty girl.

In sentences like this you must put **quite** in front of **a**. Don't say, for example, '~~She was a quite pretty girl~~'.

Don't use 'quite' in front of comparative adjectives or adverbs. Don't say, for example, '~~The train is quite quicker than the bus~~'. Use **a bit** or **slightly**.

I need something slightly cheaper.
He could run a bit more quickly than the other two men.

1 used for emphasis

Quite can be used in front of an adjective or adverb to emphasize something.

You're quite right.
I saw the driver quite clearly.

R

raise

→ see **bring up – raise – educate**

rather

1 used as adverb

Rather means 'more than a little'.

I'm rather busy at the moment.
He did it rather badly.

2 'would rather'

If you say that you **would rather do** something, you mean that you would prefer to do it.

I would rather stay in bed.

> In sentences like this you use an infinitive without 'to' after **would rather**. Don't say, for example, '~~I would rather to stay in bed~~.'

3 'rather than'

Rather than means 'instead of'. You can use **rather than** to link words, phrases or clauses.

It made him frightened rather than angry.
I use the bike when I can, rather than the car.

reach

→ see **arrive – reach – get to**

read

1 reading to yourself

When you **read** /riːd/ a piece of writing, you look at it and understand what it says.

You should read this book.

The past tense form and past participle of **read** is *read* /red/, not 'readed'.

Have you read that article I gave you?

2 reading to someone else

If you **read** something such as a book to someone, you say the words aloud so that the other person can hear them. You can either say that you read someone something, or that you read something **to** someone.

I'm going to read him some of my poems.
You should read books to your baby.

You can also say that you **read to** someone without saying what you read.

I'll go up and read to Sam for five minutes.

ready

If you get **ready**, you prepare yourself for something.

We got ready for bed.

If something is **ready**, it has been prepared and you can use it.

Lunch is ready.

> You cannot use **ready** with either of these meanings in front of a noun.

realize

→ see **understand – realize**

really

1 used for emphasis

You use **really** in conversation to emphasize something that you are saying.

Really goes in front of a verb, adjective or adverb to mean 'very' or 'very much'.

I really enjoyed that.
It was really good.
He did it really carefully.

You can put **really** in front of or after an auxiliary verb. For example, you can say 'He **really is** coming' or 'He **is really** coming'.

We really are expecting her book to do well.
It would really be too much trouble.

In formal writing, use **very** or **extremely** instead of 'really'.

2 used to show surprise

You can use **Really?** to show that you are surprised by something that someone has said.

'I think he likes you.' – 'Really? He hardly spoke to me all day.'

receipt – recipe

1 'receipt'

A **receipt** /rɪsiːt/ is a piece of paper that shows that you have received money or goods from someone.

We've got receipts for everything we've bought.

2 'recipe'

Don't confuse **receipt** with **recipe** /resəpi/. A **recipe** is a set of instructions for how to cook something.

She gave me a recipe for carrot soup.

recipe

→ see **receipt – recipe**

recommend

If you **recommend** someone or something, you praise them and advise other people to use them or buy them.

I asked my friends to recommend a doctor who is good with children.

You can say that you **recommend** someone or something **for** a particular job or purpose.

I'll recommend you for the job.

If you **recommend** something or **recommend doing** something, you say that it is the best thing to do.

The committee recommended several changes.
The doctor may recommend eating less salt.

You can recommend **that someone does** something.

We recommend that you pay in advance.

> Don't say that you 'recommend someone' an action. Don't say, for example, '~~I recommend you a visit to Paris~~'. Say 'I **recommend a visit** to Paris', 'I **recommend visiting** Paris', or 'I **recommend that you visit** Paris'.

refuse

→ see **disagree – refuse**

relation – relative – relationship

1 'relation' and 'relative'

Your **relations** or **relatives** are the members of your family.

I am a relation of her first husband.
We went to France to visit some of our relatives.

The **relations** between people or groups are the way they behave towards each other and feel about each other.

The country has good relations with Israel.

2 'relationship'

You can talk in a similar way about the **relationship** between two people or groups.

We have a good relationship with our customers.

A **relationship** is also a close friendship between two people, especially a romantic friendship.

Their relationship ended two months ago.

relationship

→ see **relation – relative – relationship**

relative

→ see **relation – relative – relationship**

relax

When you **relax**, you make yourself calmer and less worried.

Some people can't even relax when they are at home.

Relax is not a reflexive verb. Don't say that you '~~relax yourself~~'.

remain – stay

To **remain** or **stay** in a particular state means to continue to be in that state. **Remain** is more formal than **stay**.

Oliver remained silent.
I stayed awake.

If you **remain** or **stay** in a place, you do not leave it.

I was allowed to remain at home.
Fewer women these days stay at home to look after their children.

If you **stay** in a place, you live there for a short time. Don't use 'remain' with this meaning.

We stayed in Brussels for two weeks.

remark

→ see **comment – mention – remark**

remember – remind

1 'remember'

If you **remember** people or events from the past, you still have an idea of them in your mind.

He remembered the man well.

Don't use **remember** in progressive forms. Don't say, for example, '~~He is remembering the man well~~'.

If you still have an idea in your mind of something you did in the past, you can say that you **remember doing** it.

I remember asking one of my sons about this.

If you do something that you had intended to do, you can say that you **remember to do** it.

He remembered to turn the gas off.

2 'remind'

If something **reminds** you **of** something that happened in the past, it makes you think about it.

Even small objects can remind us of events in the past.

You also use **remind** to mean that you mention something that someone needs to do, so that they do not forget to do it. In this meaning, you say you **remind** someone **about** something or **remind** them **to do** something.

I need to remind people about their reports.
Remind me to speak to you about Davis.

You can also **remind** someone **that** something is the case.

I reminded him that we had a wedding to go to on Saturday.

remind

→ see **remember – remind**

remove – move

1 'remove'

If you **remove** something, you take it away.

The waiters came in to remove the cups.

2 'move'

If you go to live in a different house, don't say that you 'remove'. Say that you **move**.

Last year my parents moved from Hyde to Stepney.

rent

→ see **hire – rent – let**

responsible

1 'responsible for'

If you are **responsible for doing** something, it is your job or duty to do it.

The children are responsible for cleaning their own rooms.

If you are **responsible for** something bad that has happened, it is your fault.

They were responsible for the death of two policemen.

2 used after a noun

Responsible can also be used after a noun. If you talk about 'the person **responsible**', you mean 'the person who is responsible for what has happened'.

I hope they find the man responsible.

3 used in front of a noun

However, if you use **responsible** in front of a noun, it has a completely different meaning. It means 'sensible and showing good judgement.'

They are responsible members of the local community.
I thought it was a very responsible decision.

return

1 going back

When someone **returns** to a place, they go back there after they have been somewhere else.

I returned to my hotel.

> Don't say '~~I returned back to my hotel~~'.

Return is a fairly formal word. In conversation, you usually use **go back**, **come back**, or **get back**.

I went back to the kitchen and poured my coffee.
I have just come back from a holiday in Scotland.
I must get back to London.

2 giving or putting something back

When someone **returns** something they have taken or borrowed, they give it back or put it back.

He borrowed my best suit and didn't return it.
We returned the books to the shelf.

ride

1 'ride'

When you **ride** an animal, bicycle, or motorcycle, you control it and travel on it. The past tense form of **ride** is *rode*, and the past participle is *ridden*.

They learned to ride a bike.
He was the best horse I have ever ridden.

2 'drive'

When someone controls a car, lorry, or train, don't say that they 'ride' it. Say that they **drive** it.

It was her turn to drive the car.
Dennis has never learned to drive.

risk

1 used as a noun

If there is a **risk** of something bad, there is a possibility that it will happen.

There is a risk of flooding.

2 used as a verb

If you **risk doing** something, that thing might happen as a result of something you do. For example, if you 'risk upsetting' someone, it is possible that you will upset them.

He risked breaking his leg when he jumped.

You can also say that someone **risks doing** something when they do it even though they know it might have bad results. For example, it you 'risk phoning' someone, you phone them even though you know that it might cause problems.

If you have an expensive rug, don't risk washing it yourself.

> Don't use a 'to'-infinitive with **risk**. Don't say, for example, '~~He risked to break his leg when he jumped.~~'

rob

→ see **steal – rob**

robber

→ see **thief – robber – burglar**

round

→ see **around – round – about**

S

's

1 used to form possessives

You use possessive **'s** to show that something belongs to someone or is connected to someone. You usually use possessive **'s** when you are talking about people or animals.

With a singular noun, you add **'s**.

...Ralph's voice.
...the horse's eyes.

With a plural noun that does not end in 's', you add **'s**.

...children's games.

With a plural noun that ends in 's', you add **'**.

...my colleagues' offices.

> You do not usually add **'s** to nouns that refer to things. For example, don't say '~~the building's front~~'. Say 'the front **of the building**'.
>
> *...the end of August.*

2 other uses of 's

's is also a short form of **is**.

He's a novelist.

's can also be used as a short form of **has** when **has** is an auxiliary verb.

She's gone home.

's can also be used as a short form of **us** after **let**.

Let's go outside.

→ see **let's – let us**

safe – secure

1 used to mean 'not dangerous'

If something is **safe**, it is not dangerous and is not likely to cause harm. In this meaning, **safe** can be used in front of a noun or after a verb.

The beaches are safe for children.
This is a safe place to live.

2 used to mean 'not in danger'

If someone or something is **safe**, they are not in danger or not likely to be harmed. In this meaning, **safe** cannot be used in front of a noun.

We're safe now. They've gone.

3 'secure'

If something is **secure**, it is well protected.

Each house has secure parking.
You must keep your belongings secure.

A **secure** job will not end soon.

For now, his job is secure.

If you feel **secure**, you feel safe and happy.

With her family, she felt secure and protected.

sale

1 'sale'

The **sale** of something is the act of selling it, or the occasion on which it is sold.

There has been an increase in the sale of bread making machines.
He made a lot of money from the sale of his home in Kent.

A **sale** is an event in which a shop sells things at less than their normal price.

The bookshop is having a sale.

2 'on sale'

If you buy something **on sale**, you buy it for less than its normal price.

She bought the coat on sale at a department store.

3 'for sale'

If something is **for sale**, its owner is trying to sell it.

I asked if the horse was for sale.

salute – greet

1 'salute'

When someone such as a soldier **salutes** someone, they raise their right hand to their head as a formal sign of respect.

The two guards saluted the General.

2 'greet'

Don't use 'salute' to say that someone says hello to someone else. The word you use is **greet**.

He greeted his mother with a hug.

say – tell

1 'say'

When you **say** something, you use your voice to produce words. The past tense form and past participle of **say** is *said* /sed/, not 'sayed'.

You use **say** when you are reporting exactly what someone said.

'I feel so relaxed,' she said.

You can also report what someone said without mentioning their exact words. You use **say** and a clause beginning with **that**. You can often leave out **that**.

She said (that) they were very pleased.

> Don't put a word such as 'me' or 'her' directly after **say**. For example, don't say '~~The woman said me that Tom had left.~~' Say 'The woman **said** that Tom had left.'

2 'tell'

If you are mentioning the hearer as well as the speaker, you usually use **tell**, not 'say'. The past tense form and past participle of **tell** is *told*.

He told me that he was sorry.

> Don't use 'to' after **tell**. Don't say, for example '~~He told to me that he was sorry.~~'

When you are talking about orders or instructions, you use **tell**, not 'say'. **Tell** is followed by a 'to'-infinitive.

She told me to be careful.

You say that someone **tells** a story, a lie, or a joke.

You're telling lies now.

3 'ask'

Don't say that someone 'says' a question. Say that they **ask** a question.

I wasn't the only one asking questions.

school

→ see topic: **Places**

search – look for

1 used as a verb

If you **search** a place, you examine it carefully because you are trying to find something.

Police searched the hospital yesterday.

> Don't say that you 'search' the thing you are trying to find. You can say that you **search for** it or that you **look for** it.
>
> *Police are searching for clues.*
> *He's looking for his keys.*

2 used as a noun

A **search** is an attempt to find something by looking for it carefully.

I found the keys after a long search.

secure

→ see **safe – secure**

see

The past tense form of **see** is *saw*, not 'seed'. The past participle is *seen*.

1 using your eyes

If you **can see** something, you are aware of it through your eyes.

I can see the sea.

> If you are talking about the present, you usually use **can see**. Don't say '~~I see the sea~~'. Also, don't use **see** in progressive forms. Don't say '~~I am seeing the sea~~'.

If you are talking about the past, you use **could see** or **saw**.

He could see Wilson's face in the mirror.
We suddenly saw a boat in the distance.

2 meeting someone

If you **see** someone, you visit them or meet them.

You should see a doctor.

3 understanding

See is also used to mean 'understand'.

Oh, I see what you mean.

see – look at – watch

1 'see'

When you **see** something, you are aware of it through your eyes, or you notice it. You can **see** things by chance.

She saw a man standing outside the building.

→ see **see**

2 'look at'

When you **look at** something, you deliberately direct your eyes towards it.

He looked at the food on his plate.

→ see **look**

3 'watch'

When you **watch** something, you look at it for a period of time.

I was here at home, watching TV.
More than 1.2 million people watched the match.

seem

You use **seem** to say that someone or something gives a particular impression.

1 used with adjectives

Seem is usually followed by an adjective. You can use the adjective on its own, or after **to be**.

Even minor problems seem important.
You seem to be very interested.

Some adjectives cannot be used in comparative forms, for example **dead**, **alive** and **asleep**. For these adjectives, you use **seem to be**, not 'seem'.

She seemed to be asleep.

2 used with noun phrases

You can also use a noun phrase after **seem** or **seem to be**. For example, instead of saying 'She seemed nice', you can say

'She **seemed a nice person**' or 'She **seemed to be a nice person**'. In conversation, people often say 'She **seemed like a nice person**'.

It seemed a long time before the food came.
She seems to be a very happy girl.
It seemed like a good idea.

shall

→ see **will – shall**

shave

When a man **shaves**, he cuts hair from his face using a razor.

He had a bath and shaved.

> **Shave** is not usually a reflexive verb. Don't say '~~He had a bath and shaved himself~~'.

shop – store

1 'shop' and 'store'

In British English, a building where you buy things is usually called a **shop**. In American English, it is usually called a **store**.

...a toy shop.
...a grocer's store.

In both British and American English, a large shop that has separate departments selling different types of goods is called a **department store**.

...the furniture department of a large department store.

2 'shopping'

Shopping is the activity of buying things from shops.

I don't like shopping.

When you go to the shops to buy things, you **go shopping**.

They went shopping after lunch.

When you go to the shops to buy things that you need regularly such as food, you **do the shopping**.

Who's going to do the shopping?

Shopping is an uncountable noun. Don't say, for example, '~~Who's going to do the shoppings?~~'

→ see also **magazine – shop**

short – shortly – briefly

1 'short'

Short is an adjective. If something is **short**, it does not last for a long time.

...a short holiday.

2 'shortly'

Shortly is an adverb. If something is going to happen **shortly**, it is going to happen soon.

Please take a seat. Dr. Garcia will see you shortly.

3 'briefly'

Don't use 'shortly' to say that something lasts for a short time. The word you use is **briefly**.

She told them briefly what had happened.

shortly

→ see **short – shortly – briefly**

should – ought to

1 moral rightness

You use **should** or **ought to** to say that something is the right thing to do.

Crimes should be punished.
I ought to call the police.

2 giving advice

You can say **you should** or **you ought to** when you are giving someone advice.

You should be careful.
I think you ought to see a doctor.

3 expectation

You use **should** or **ought to** to say that you expect something to happen.

We should be there by dinner time.
The party ought to be fun.

4 negative forms

Should and **ought to** have the negative forms **should not** and **ought not to**. These are often shortened to **shouldn't** and **oughtn't to**.

You shouldn't go to the meeting.
We oughtn't to laugh.

shut

→ see **close – closed – shut**

sick

→ see **ill – sick**

since – for

1 'since'

You use **since** to talk about something that started in the past, and that has continued from then until now.

I've been wearing glasses since I was three.

> In sentences like this you use a perfect form with **since**. Don't say '~~I wear glasses since I was three.~~'

2 'for'

You use **for** to say how long something lasts or continues.

I'm staying with Bob for a few days.

You also use **for** to say how long something has lasted or continued.

He has been missing for three weeks.

> When you use **for** to say how long something has lasted or continued, you must use a perfect form. You cannot say, for example, '~~He is missing for three weeks~~'.

sleep – asleep

1 'sleep'

Sleep can be a noun or a verb. The past tense form and past participle of the verb is *slept*, not 'sleeped'.

Sleep is the natural state of rest when your eyes are closed and your body is not active.

I haven't been getting enough sleep recently.

When you **sleep**, you are in this state of rest.

He was so excited he couldn't sleep.

2 'asleep'

If someone is in this state of rest, you say they are **asleep**.

She was asleep in the guest room when we walked in.

> Don't use **asleep** in front of a noun. Don't, for example, talk about an '~~asleep child~~'. Instead, you can say a '**sleeping child**'.
>
> Don't say that someone is '~~very asleep~~'. You can say that they are **sound asleep** or **fast asleep.**
>
> *Chris is still sound asleep in the other bed.*
> *Colette was fast asleep when we left.*

3 'go to sleep'

When someone changes from being awake to being asleep, you say that they **go to sleep**.

Be quiet and go to sleep!

4 'fall asleep'

When someone goes to sleep suddenly or unexpectedly, you say that they **fall asleep**.

I fell asleep during the film.

smell

The past tense form and past participle of the verb is *smelled*, but *smelt* is also used in British English.

1 used as an intransitive verb

You can say that a place or object **smells of** a particular thing.

The room smelled of roses.
Her clothes smelt of smoke.

You can also use **smell** with an adjective to say that something has a pleasant or unpleasant smell.

That soup smells delicious!

If you say that something **smells**, you mean that it has an unpleasant smell.

The fridge is beginning to smell.

2 used as a transitive verb

If you **can smell** something, you are aware of it through your nose.

I can smell the ocean.

> Don't use **smell** in progressive forms. Don't say '~~I am smelling the ocean.~~'

If you are talking about the past, you use **could smell**, **smelled** or **smelt**.

I could smell coffee.
People said they smelled gas in the building.

so

1 referring back

You often use **so** after verbs like **think**, **hope**, **expect** and **say**. For example, if someone says 'Is Alice at home?', you can say 'I **think so**', meaning 'I think Alice is at home'.

'Are you all right?' – 'I think so.'
'Will you be able to take driving lessons at your new school?' – 'I expect so.'

2 used for emphasis

You can use **so** to emphasize an adjective.

These games are so boring.

However, if the adjective is in front of a noun, you use **such a**, not 'so'.

She was so nice.
She was such a nice girl.

→ see **such**

You can also use **so** to emphasize an adverb.

She sings so beautifully.

so – very – too

So, **very**, and **too** can all be used to emphasize the meaning of an adjective, an adverb, or a word like **much** or **many**.

1 'very'

Very is the simplest way to make something stronger. It has no other meaning.

I'm very happy.

2 'so'

You can use **so** to show that you feel strongly about something.

John makes me so angry!
Oh thank you so much!

3 'too'

You use **too** to talk about something that is more than you want or need.

She wears too much make-up.
Sorry, I can't stay. I'm too busy.

some – any

1 'some'

You use **some** to talk about an amount of people or things, without saying exactly how many or how much. You can use **some** in front of a plural or an uncountable noun.

We saw some children in the park.
She had a piece of cake and some coffee.

When you use **some** in front of a plural noun, you use a plural form of a verb.

Here are some suggestions.

When you use **some** in front of an uncountable noun, you use a singular form of a verb.

There is some rice in the fridge.

2 'some of'

You use **some of** in front of a singular or plural noun phrase beginning with a determiner (a word such as **the**, **these**, **my** or **his**).

I have read some of his stories.
They took some of the money away.

You can also use **some of** in front of a plural pronoun such as **us** or **them**.

Some of them have young children.

3 'any' used in negative sentences

Don't use 'some' in negative sentences. Use **any**. You can use **any** in front of a plural or uncountable noun.

I don't have any plans for the summer holidays.
We made this without any help.

4 questions

You usually use **any** in questions.

Do you have any questions?
Were you in any danger?

However, if you expect the answer to be 'yes', you can use **some**.

Have you lost some weight?

You also use **some** when you are offering something.

Would you like some cake?

somebody

→ see **someone – somebody – anyone – anybody**
→ see topic: **Talking about men and women**

someone – somebody – anyone – anybody

Someone and **anyone** are more common in written English. **Somebody** and **anybody** are more common in conversation.

1 statements

You use **someone** or **somebody** to talk about a person without saying who you mean.

Carson sent someone to see me.
There was an accident and somebody was killed.

2 negative sentences

> Don't use 'someone' or 'somebody' in negative sentences. Use **anyone** or **anybody**.
>
> *There wasn't anyone there.*

3 questions

You usually use **anyone** or **anybody** in questions.

Was anybody in the office?

However, if you expect the answer to be 'yes', you can use **someone** or **somebody**.

Did you meet someone last night?

> Don't use 'someone' or 'somebody' with 'of' in front of a plural noun. Don't say, for example, '~~Someone of my friends is a painter~~'. Say '**One of** my friends is a painter'.

4 'some people'

Someone and **somebody** do not have plural forms. If you want to talk about a group of people without saying who you mean, you say **some people**.

Some people attempted to walk across the bridge.

→ see also topic: **Talking about men and women**

something – anything

1 statements

You use **something** to talk about an object or situation without saying exactly what it is.

Hendricks saw something ahead of him.
It's something that has often puzzled me.

2 negative sentences

Don't use 'something' in negative sentences. Instead, use **anything**.

I did not say anything.
He never seemed to do anything at all.

3 questions

You usually use **anything** in questions.

Did you buy anything?

However, if you expect the answer to be 'yes', you can use **something**.

Has something happened?

You can also use **something** when you are offering something.

Would you like something to eat?

sometime

→ see **sometimes – sometime**

sometimes – sometime

1 'sometimes'

You use **sometimes** to say that something happens on some occasions, but not all the time.

Sometimes I wish I was back in Africa.

2 'sometime'

Sometime or **some time** means at a time in the past or future that you do not know yet.

Can I come and see you sometime?
He died some time last year.

somewhere – anywhere

1 statements

You use **somewhere** to talk about a place without saying exactly where you mean.

They lived somewhere near Bournemouth.

2 negative sentences

Don't use 'somewhere' in negative sentences. Use **anywhere**.

I haven't got anywhere to live.

3 questions

You usually use **anywhere** in questions.

Is there an ashtray anywhere?

However, if you expect the answer to be 'yes', you can use **somewhere**.

Are you taking a trip somewhere?

Some American speakers say **someplace** or **some place** instead of 'somewhere'.

She had seen it someplace before.
Why don't we go some place where it's quieter?

sound

1 used as a verb

You use **sound** in front of an adjective when you are describing something that you hear.

Her voice sounded strange.

You use **sound** in front of an adjective to describe how someone seems when they speak.

He sounded angry.

You use **sound** in front of an adjective to describe your opinion of something.

'They've got a small farm in Devon.' – 'That sounds nice.'

> Don't use **sound** in progressive forms. Don't say, for example, '~~That is sounding nice~~'.

2 'sound like'

You use **sound like** and a noun phrase to say that something has a similar sound to something else.

Her footsteps sounded like gun shots.

You use **sound like** and a noun phrase to describe your opinion of something.

That sounds like a good idea.

sound – noise

1 used as countable nouns

A **sound** is something that you can hear. If it is unpleasant or unexpected, you call it a **noise**.

We listened to the sounds of the forest.
A sudden noise made Brody jump.

2 used as uncountable nouns

Sound is the general term for things that can be heard. However, after expressions such as **too much** or **a lot of**, or if it is unpleasant, you use **noise**.

...the speed of sound.
Try not to make so much noise.

south

→ see topic: **North, South, East and West**

southern

→ see topic: **North, South, East and West**

souvenir – memory

1 'souvenir'

A **souvenir** /suːvəniə/ is an object that you buy or keep to remind you of a holiday, place, or event.

He kept the spoon as a souvenir of his journey.

2 'memory'

Don't use 'souvenir' to talk about something that you remember. The word you use is **memory**.

Her earliest memory is of singing to her family.

Your **memory** is your ability to remember things.

You've got such a wonderful memory.

speak – talk

1 'speaking' and 'talking'

When you mention that someone is using his or her voice to produce words, you say that they **are speaking**.

He didn't look at me while I was speaking.

However, if two or more people are having a conversation, you say that they **are talking**.

They sat in the kitchen drinking tea and talking.

2 used with 'to' and 'with'

If you **speak to** someone or **talk to** them, you have a conversation with them.

I saw you speaking to him just now.
I enjoyed talking to Anne.

Some American speakers say **speak with** or **talk with**.

When he spoke with his friends, he told them what had happened.

When you make a telephone call, you ask if you can **speak to** someone. Don't ask if you can 'talk to' them.

Hello. Could I speak to Sue, please?

3 languages

You say that someone **speaks** or **can speak** a foreign language. Don't say that someone 'talks' a foreign language.

They spoke fluent English.

If you say, for example, 'Kate **speaks** Dutch', you mean that she is able to speak Dutch. If you say 'She **is speaking** Dutch' or 'She **is speaking in** Dutch', you mean that she is speaking Dutch at the moment.

spend – pass

1 'spend'

If someone does something from the beginning to the end of a period of time, you say that they **spend** the time doing it.

We spent the evening talking about art and the theatre.

You can say that someone **spends** a period of time with someone or in a particular place.

I spent an evening with Davis.
He spent most of his time in the library.

2 'to pass the time'

If you do something to occupy yourself while you are waiting for something, you say that you do it **to pass the time**.

He brought a book along to pass the time.

3 'have'

If you enjoy yourself while you are doing something, don't say that you 'pass' or 'spend' a good time. Say that you **have** a good time.

We had a marvellous time in Spain.

spring

→ see topic: **Seasons**

staff

The people who work for an organization are its **staff**.

She joined the staff of the BBC.

You usually use a plural form of a verb after **staff**.

The staff are very helpful.

If you want to talk about one person who works for an organization, you say a **member of staff**.

I'd like to speak to a member of staff.

start – begin

1 used with noun phrases

If you **start** or **begin** something, you do it from a particular time.

My father started work when he was ten.
The US is prepared to begin talks immediately.

The past tense form of **begin** is *began*, and the past participle is *begun*.

2 used with other verbs

You can use a 'to'-infinitive or an '-ing' form after **start** and **begin**.

Ralph started to run.
He started laughing.

If you use an '-ing' form of **start** or **begin**, the next verb must be a 'to'-infinitive.

Now that I'm better, I'm beginning to eat more.

3 used as intransitive verbs

If something **starts** or **begins**, it happens from a particular time.

His meeting starts at 7.
My career as a journalist was about to begin.

stay

→ see **remain – stay**

steal – rob

1 'steal'

When someone **steals** something, they take it without permission and without intending to return it.

He tried to steal a car from a car park.

The past tense form of **steal** is *stole* and the past participle is *stolen*.

2 'rob'

When you are speaking about the thing that was stolen, you use **steal**. However, when you are speaking about the person or place it was taken from, you use **rob**.

They robbed me and stole my car.
He was arrested for robbing a bank.

still

1 meaning 'continuing'

Still is used to say that a situation continues to exist or that something continues to happen.

She still lives in London.

2 used in negative statements

You can use **still** in a negative statement, for emphasis. **Still** goes in front of the auxiliary verb.

I still don't understand.
My father still hasn't forgiven me.

Don't use 'still' to say that something has not happened up to the present time. The word you use is **yet**.

It isn't dark yet.

→ see **yet**

→ see also topic: **Where you put adverbs**

sting – bite

1 'sting'

The past tense form and past participle of **sting** is *stung*.

If an insect **stings** you, it pushes a poisoned part into your skin.

He was stung inside his mouth by a wasp.

2 'bite'

Don't say that a mosquito or ant 'stings' you. Say that it **bites** you. The past tense form of **bite** is *bit* and the past participle is *bitten*.

A mosquito bit her on the wrist.

You also say that a snake **bites** you.

He was bitten by a snake.

stop

1 'stop doing'

If you **stop doing** something at a particular time, you no longer do it after that time.

We all stopped talking.

2 'stop to do'

If you **stop to do** something, you interrupt what you are doing in order to do something else.

I stopped to tie my shoelace.

store

→ see **shop – store**

storey

→ see **floor – storey – ground**

stranger

A **stranger** is someone who you have never met before.

I was in a room full of complete strangers.

> Don't use 'stranger' to talk about someone who comes from a country that is not your own. You can say **foreigner**, but this word can sound rude. It is better to talk about 'someone **from abroad**'.
>
> *...visitors from abroad.*

street

A **street** is a road in a town or a large village, usually with buildings along it.

The two men walked slowly down the street.

> Don't use 'street' to talk about a road in the countryside. Instead, use **road**.

such

1 'such as'

You use **such as** to give an example of something.

The hotel provides facilities such as the internet.

2 'such' used for emphasis

Such is used to emphasize the adjective in a noun phrase.

She was such a nice girl.

> You must use **a** when the noun phrase is singular. Don't say, for example, '~~She was such nice girl~~'.

suggest

When you **suggest** something, you mention it as a plan or idea for someone to think about.

We have to suggest a list of topics for next term's lessons.

You can **suggest** that someone does something.

I suggest you ask him some questions about his past.

> Don't use the name of a person directly after **suggest**. If you want to say who the suggestion is for, use **to**. Don't say, for example '~~John suggested me the idea.~~' Say 'John suggested the idea **to me**.'

suit

→ see **fit – suit**

summer

→ see topic: **Seasons**

supper

→ see topic: **Meals**

support

If you **support** someone or **support** their aims, you agree with them and try to help them.

They have supported our efforts to raise money for a new gym.

If you **support** a sports team, you want them to win.

He has supported Arsenal all his life.

If you **support** someone, you give them money or the things they need.

He has to support his wife and children.

Don't use 'support' to say that someone accepts pain or an unpleasant situation. Say that they **bear** it or **put up with** it.

If you do not like something at all, don't say that you '~~can't support~~' it. Say that you **can't bear** it or **can't stand** it.

→ see **bear – can't stand – put up with**

suppose – assume

1 'suppose'

If you **suppose** that something is true, you think it is probably true.

I suppose you're right.

2 'assume'

If you **assume** that something is true, you are fairly sure about it, and act as if it were true.

I assumed that he was talking about his wife.

3 'be supposed to'

If something **is supposed to** be done, it should be done.

You are supposed to report it to the police as soon as possible.
I'm not supposed to talk to you about this.

If something **is supposed to** be true, people think that it is true.

Swimming <u>is supposed to</u> be very good exercise for older people.

> Remember the 'd' on **supposed**. Don't say '~~is suppose to~~.'

surely – definitely – certainly

1 'surely'

You use **surely** for emphasis when you think that something should be true.

<u>Surely</u> they don't all work here.
<u>Surely</u> they could have done something to help her.

2 'definitely' and 'certainly'

When you use 'surely', there may be some doubt. If there is no doubt at all about something, you use **definitely** or **certainly**.

They were <u>definitely</u> not for sale.
If nothing is done, there will <u>certainly</u> be an economic crisis.

In British English, you do not use 'surely' as a way of agreeing with someone or saying 'yes'. The word you use is **certainly**.

'Isn't it ugly?' – 'It <u>certainly</u> is!'
'Can you arrange an early morning call, please?' – 'Yes, <u>certainly</u>.'

American speakers use both **surely** and **certainly** to agree or say 'yes'.

surprise

1 used as a verb

If something **surprises** you, you did not expect it.

Dad's reply surprised me.

> Don't use a progressive form of **surprise**. Don't say, for example, '~~Dad's reply was surprising me~~'.

2 used as a noun

If something is a **surprise**, it was not expected.

It was a great surprise to find out I had won something.

3 'surprised'

Surprised is an adjective that is followed by a 'to'-infinitive. For example, if you are **surprised to see** something, you did not expect to see it.

You'll be surprised to learn that Charles Boon is living here.

> Remember the 'd' on **surprised**. Don't say that someone is '~~surprise to~~' see or hear something.

sympathetic – nice – likeable

1 'sympathetic'

If someone is being **sympathetic**, they are kind to someone who has problems, and show that they understand their feelings.

My boyfriend was very sympathetic and it made me feel better.

2 'nice' and 'likeable'

Don't say that someone is 'sympathetic' when they are pleasant and easy to like. The word you use is **nice** or **likeable**.

He was a terribly nice man.
...a very likeable and attractive young woman.

T

take

The other forms of **take** are *takes*, *taking*, *took*, *taken*.

1 actions and activities

You use **take** with a noun to talk about an action.

She took a shower.
We took a walk to the park.

With some nouns such as 'bath' and 'shower', you can use **take** or **have** with no difference in meaning.

→ see **have**

2 moving things

If you **take** something from one place to another, you carry it there.

Don't forget to take your umbrella.

→ see **carry – take**
→ see also **bring – take – fetch**

3 exams and tests

When someone completes an exam or test, you say that they **take** the exam or test.

She took her driving test last year.

4 time

If something **takes** a certain amount of time, you need that much time to do it.

It may take them several weeks to get back.

→ see also topic: **Transport**

take place

→ see **happen – take place – occur**

talk

→ see **speak – talk**

tall

→ see **high – tall**

tea

→ see topic: **Meals**

teach

→ see **learn – teach**

teacher

→ see **professor – teacher**

terrible – terribly

1 'terrible'

In conversation, **terrible** means 'extremely bad'.

His eyesight is terrible.

In writing or conversation, **terrible** means 'shocking or upsetting'.

There was a terrible plane crash last week.

2 'terribly'

You can use the adverb **terribly** to emphasize how bad something is.

Our team played terribly today.

In conversation, you can use **terribly** to emphasize a verb or an adjective.

I'm terribly sorry.
We all miss him terribly.

Don't use **terribly** like this in formal writing.

test

→ see **exam – test**
→ see **prove – test**

than

1 'than' used with comparatives

You mainly use **than** when you use comparative adjectives and adverbs.

I am happier than I used to be.
They had to work harder than the younger boys.

If you use a personal pronoun on its own after **than**, it must be an object pronoun such as **me** or **him**.

My brother is younger than me.

However, if the pronoun is followed by a verb, you use a subject pronoun such as **I** or **he**.

He's taller than I am.

2 'more than'

You use **more than** to talk about a greater number of people or things.

... a city of more than a million people.

→ see **more**

that – those – this – these

These is the plural form of **this**. **Those** is the plural form of **that**.

1 'that' and 'those' used to talk about things that have been mentioned

You can use **that** or **those** to talk about people, things, or events that have already been mentioned.

I knew that meeting would be difficult.
Not all crimes are committed for those reasons.

2 'that' and 'those' used for things you can see

You can also use **that** or **those** to talk about people or things that you can see but that are not close to you.

Look at that bird!
Don't be afraid of those people.

3 'this' and 'these' used to talk about things that have been mentioned

You can use **this** or **these** to talk about people, things, or events that have just been mentioned.

Tax increases may be needed next year to do this.
These particular students are extremely clever.

4 'this and those' used for things near you

You can use **this** or **these** to talk about people or things that are near you.

This book is very good.
I'm sure they don't have chairs like these.

You use **this is** when you are introducing someone.

This is Bernadette, Mr Zapp.

You also use **this is** to say who you are when you phone someone.

Sally? This is Martin Brody.

the

1 basic uses

You use **the** before a noun when it is clear which person or thing you are talking about.

The doctor will be here in a minute.

You use **the** before a singular noun when there is only one such thing.

They all sat in the sun.

2 types of thing or person

Don't use 'the' with plural or uncountable nouns when you are talking about things in general. For example, if you are talking about pollution in general, say '**Pollution** is a serious problem'. Don't say '~~The pollution is a serious problem~~'.

...victims of crime.
We all know computers can make mistakes.

You can use **the** with words such as **rich**, **poor**, **young**, **old**, or **unemployed** to talk about all people of a particular type.

They were discussing the problem of the unemployed.

When you use one of these words like this, don't add '-s' or '-es' to it. Don't talk, for example, about '~~the unemployeds~~'.

3 nationalities

You can use **the** with some nationality adjectives to talk about the people who come from a particular country.

The French and the British will not agree on this.

4 musical instruments

You usually use **the** with the name of a musical instrument when you are talking about someone playing it:

You play the guitar, I see.

5 used with superlatives and comparatives

You usually use **the** before superlative adjectives.

...the smallest church in England.

You don't usually use 'the' in front of superlative adverbs.

...the language they know best.

You don't usually use 'the' in front of comparative adjectives or adverbs.

I wish we could do this quicker.

→ see also topic: **Places**
→ see also topic: **Meals**

there

1 'there is' and 'there are'

You use **there is**, **there are**, **there was** or **there were** to say that something exists or happens, or that something is in a particular place. When you use **there** like this, you pronounce it /ðe/ or /ðə/.

You use **there is** or **there was** with a singular noun phrase, and **there are** or **there were** with a plural noun phrase.

There was a fire on the fourth floor.
Are there any biscuits left?

Don't use 'there is' or 'there are' with 'since' to say how long ago something happened. Don't say, for example, '~~There are four days since she arrived in London~~'. Say '**It's** four days since she arrived in London' or 'She arrived in London four days **ago**'.

→ see **since – for**

2 used as an adverb

You can also use **there** to talk about a place that has just been mentioned. When you use **there** like this, you pronounce it /ðeə/.

I must get home. Bill's there on his own.

Don't use 'to' in front of **there**. Don't say, for example, '~~I like going to there~~'. Say 'I like going **there**'.

My family live in India. I still go there often.

3 'their'

Don't confuse **there** with **their**, which is also pronounced /ðeə/. Use **their** to show that something belongs to particular people, animals, or things.

I looked at their faces.

the rest

The rest of something means 'the remaining parts of something'.

I ate two cakes and saved the rest.
I'll remember that experience for the rest of my life.

If you use **the rest of** followed by an uncountable noun, you use a singular verb.

The rest of the food was delicious.

If you use **the rest of** followed by a plural noun, you use a plural verb.

The rest of the boys were delighted.

these

→ see **that – those – this – these**

they

→ see topic: **Talking about men and women**

thief – robber – burglar

A **thief** is someone who steals. A **robber** often uses violence to steal things from places such as banks or shops.

They were attacked by a group of robbers.

A **burglar** breaks into buildings and steals things.

Most burglars spend just two minutes inside a house.

think

The past tense form and past participle of **think** is *thought*, not 'thinked'.

1 giving an opinion

You can use **think** when you are giving your opinion about something.

I think you should go.

> When you use **think** like this, don't use a progressive form. Don't say, for example, '~~I am thinking you should go~~'.

Instead of saying that you think something is not true, you usually say that you **don't think** it is true.

I don't think there is any doubt about that.

2 'I think so'

If someone asks you if something is true, you can say '**I think so**'. Don't say '~~I think it~~'.

'Is he still in Sydney?' – 'I think so.'

If you want to reply that something is probably not true, you say '**I don't think so**'.

'Are you going to be sick?' – 'I don't think so.'

3 using a progressive form

When someone **is thinking**, they are using their mind to consider something. When you use **think** with this meaning, you often use a progressive form. You can say that someone **is thinking about** something or someone, or **is thinking of** something or someone.

I spent hours thinking about the interview.
She was thinking of her husband.

If you are considering doing something, you can say that you **are thinking of doing** it.

I 'm thinking of going to college next year.

Don't say '~~I'm thinking to go to college next year.~~'

this

→ see **that – those – this – these**
→ see topic: **Times of the day**

those

→ see **that – those – this – these**

thousand

→ see **hundred – thousand – million**

time

1 'time'

Time is something that we measure in hours, days, years, etc.

...a period of time.
More time passed.

> You don't usually use 'time' when you are saying how long something takes or lasts. Don't say, for example, '~~Each song lasts ten minutes' time~~'. Say 'Each song lasts **ten minutes**'.
>
> *The whole process takes twenty-five years.*

However, you can use **time** when you are saying that something will happen in the future.

We are getting married in two years' time.

> **Time** is usually an uncountable noun, so don't use 'a' with it. Don't say, for example, '~~I haven't got a time to go shopping~~'. Say 'I haven't got **time** to go shopping'.
>
> *Have you got time for tea?*

2 'a...time'

However, you can use **a** with an adjective and **time** to say how long something takes or lasts. You can say, for example, that something takes **a long time** or takes **a short time**.

The storm lasted a long time.

If you are enjoying yourself, you can say that you **are having a good time**.

Did you have a good time in Edinburgh?

You must use **a** in sentences like these. Don't say, for example, '~~Did you have good time in Edinburgh?~~'.

3 'on time'

If something happens **on time**, it happens at the right time.

The train arrived at the station on time.

4 'in time'

Don't confuse **on time** with **in time**. If you are **in time** for something, you are not late for it.

We're just in time.
He returned to his hotel in time for dinner.

today

Today means 'this day' or 'on this day'.

I had a letter today from my sister.
Today is Thursday.

Don't use 'today' in front of **morning**, **afternoon**, or **evening**. Instead, you use **this**.

His plane left this morning.

→ see also **topic: Times of the day**

tomorrow → see topic: **Times of the day**

tonight → see topic: **Times of the day**

too

→ see **also – too – as well**
→ see **so – very – too**

traffic

You use **traffic** to talk about all the vehicles that are on a particular road at one time.

There was heavy traffic on the road.

> **Traffic** is an uncountable noun. Don't talk about '~~traffics~~' or '~~a traffic~~'.

travel

Travel can be a verb or a noun. The other forms of the verb are *travels, travelling, travelled* in British English, and *travels, traveling, traveled* in American English.

1 used as a verb

If you make a journey to a place, you can say that you **travel** there.

I travelled to work by train.

When you **travel**, you go to several places, especially in foreign countries.

You need a passport to travel abroad.

2 used as a noun

Travel is the act of travelling. It is usually an uncountable noun.

He hated air travel.

Don't talk about 'a travel'. Instead talk about a **journey**, a **trip**, or a **voyage**.

→ see **journey – trip – voyage**

trip

→ see **journey – trip – voyage**

trouble

1 used as an uncountable noun

Trouble is usually an uncountable noun. You use **trouble** to talk about different kinds of problems or difficulties.

You've caused us a lot of trouble.

You can say that someone **has trouble doing** something.

Did you have any trouble finding your way here?

Don't say '~~Did you have any trouble to find your way here?~~'.

2 'troubles'

Your **troubles** are the problems in your life.

It helps me forget my troubles and relax.

Don't call a single problem 'a trouble'. Use **problem**.
Pollution is a problem in this city.

3 'the trouble'

You use **the trouble** to talk about a particularly difficult part of a problem.

It's getting a bit expensive now, that's the trouble.

trousers

Trousers are a piece of clothing that covers your body from the waist downwards, and covers each leg separately. **Trousers** is a plural noun. You use a plural form of a verb with it.

His trousers were covered in mud.

> Don't talk about '~~a trousers~~'. Say **some trousers** or **a pair of trousers**.
>
> *Claud was wearing a pair of black trousers.*

You use a singular form of a verb with **a pair of trousers**.

There was a pair of trousers on the bed.

In American English, people don't usually say **trousers**. They say **pants** or **slacks**.

true – come true

1 'true'

A **true** story is based on facts, and is not invented or imagined.

The story about the murder is true.

2 'come true'

If a dream **comes true**, it actually happens.

Some dreams come true.

> Don't use 'become'. Don't say, for example, '~~Some dreams become true.~~'

try – attempt

The other forms of **try** are *tries, trying, tried.*

1 'try'

If you **try to do** something, you make an effort to do it.

He was trying to understand.

You **try doing** something in order to find out what it is like.

Why don't you try cooking it in the oven first.

2 'attempt'

If you **attempt to do** something, you try to do it. **Attempt** is a more formal word than **try**.

Now we'll attempt to have a conversation.

> You usually use a 'to'-infinitive after **attempt**. Don't say, for example, '~~Now we'll attempt having a conversation.~~'

U

understand – realize

1 'understand'

If you can **understand** someone, you know what they mean.

His lecture was confusing; no one could understand it.

If you say that you **understand** that something is true, you mean that you have been told that it is true.

I understand he's been married before.

2 'realize'

Don't use 'understand' to say that someone becomes aware of something. Use **realize**.

As soon as I saw him, I realized that I'd seen him before.

university → see topic: **Places**

unless

You use **unless** to say that something can only happen or be true if something else happens or is true. For example, if you say 'I won't go to the party unless you go with me', you mean 'I'll go to the party if you go with me'.

We won't use force unless we have to.

> Use the present simple after **unless**. Don't say, for example, '~~We won't use force unless we will have to~~'.

When you are talking about a situation in the past, use the past simple after **unless**.

She wouldn't go with him unless I came too.

used to

1 talking about the past

If something **used to** /juːs tuː, juːs tə/ be true, it was true in the past but is not true now.

She used to live in Egypt.
I used to be afraid of you.

2 'didn't use to'

In conversation, you can say that something **didn't use to** be true.

The house didn't use to be so clean.

3 familiarity

If you are **used to** something, you are familiar with it and you accept it. With this meaning, **used to** comes after the verb **be**, and is followed by a noun phrase or an '-ing' form.

The noise doesn't frighten them. They're used to it.
I'm used to getting up early.

usual – usually

1 'usual'

Usual is used to describe the thing that happens, is done or is used most often. **Usual** normally comes after **the** or a word such as **his** or **my**. Don't use it after 'a'.

They are not taking the usual amount of exercise.
He sat in his usual chair.

If you want to compare a situation with another situation that happens more often, you can use a comparative adjective followed by **than usual**, or a verb followed by **more than usual**.

February was colder than usual.
They complained more than usual.

You can say that it is **usual for** someone **to do** something.

It is usual for them to meet regularly.

2 'ordinary'

Don't use 'usual' to say something is not of a special kind. Use **ordinary**.

These children should be educated in an ordinary school.

3 'usually'

You use the adverb **usually** to talk about the thing that most often happens.

We usually eat in the kitchen.

4 'as usual'

When something that happens is the thing that most often happens, you can say that it happens **as usual**.

She wore, as usual, her black dress.

usually

→ see **usual – usually**

V

vacation

→ see **holiday – vacation**

very

1 basic use

You use **very** to emphasize an adjective or adverb.

...a very small child.
Joe was very worried about her.
Think very carefully.

2 when you cannot use 'very'

Don't say that someone is '~~very awake~~' or '~~very asleep~~', or that two things are '~~very apart~~'. Say that they are **wide awake**, **fast asleep**, or **far apart**.

He was wide awake all night.
Chris is still fast asleep in the other bed.
His two hands were far apart.

Don't use 'very' with adjectives that already describe an extreme quality. Don't say, for example, that something is '~~very enormous~~'. Here is a list of adjectives that you cannot use with 'very':

- awful
- brilliant
- enormous
- excellent
- huge
- terrible
- wonderful

3 comparatives and superlatives

Don't use 'very' with comparatives. Don't say, for example, '~~Tom was very quicker than I was~~'. Say 'Tom was **much quicker** than I was' or 'Tom was **far quicker** than I was'.

It is much colder than yesterday.
It is a far better picture than the other one.

You can use **very** in front of **best**, **worst**. or any superlative which ends in '-est'.

It's one of Shaw's very best plays.
...the very latest photographs.

→ see also **so – very – too**

view

→ see **point of view – view – opinion**

visit

1 used as a verb

If you **visit** a place, you go to see it because you are interested in it.

He'll visit four cities on his trip.

If you **visit** someone, you go to see them or stay with them at their home.

She visited some of her relatives for a few days.

2 used as a noun

Visit is also a noun. You can **make** a visit to a place or **pay** a visit to someone.

He made a visit to the prison that day.
It was too late to pay a visit to Sally.

> Don't use 'do'. Don't say, for example, '~~It was too late to do a visit to Sally~~'.

voyage

→ see **journey – trip – voyage**

W

wait

1 'wait'

When you **wait**, you spend time doing very little, until something happens or someone arrives.

She was waiting in the queue to buy some stamps.

2 'wait for'

You can say that someone **waits for** something or someone.

I'm waiting for Joan.

You can also say that someone **waits for** a person or thing **to do** something.

She waited for me to say something.

> **Wait** is never a transitive verb. Don't say, for example, '~~I'm waiting Joan.~~' You must use **wait for**.

want

1 basic use

If you **want** something, you feel a need for it.

Do you want a cup of coffee?

> Don't use a progressive form of **want**. Don't say, for example, '~~Are you wanting a cup of coffee?~~'

2 used with a 'to'-infinitive

You can say that someone **wants to do** something.

They wanted to go shopping.

> Don't say that someone 'wants to not do' something or 'wants not to do' something. Say that they **don't want to do** it.
>
> *I don't want to discuss this.*

You can say that you **want** someone **to do** something.

I want him to learn to read.

> Don't use 'that' after **want**. Don't say, for example, '~~I want that he should learn to read~~'.

3 requests

> You don't normally use 'want' when you are making a request. Don't walk into a shop and say 'I want a box of matches, please'. Say 'Could I have a box of matches, please?'

wash

1 used with an object

If you **wash** something, you clean it with water and soap.

She washes and irons his clothes.

You can **wash** a part of your body.

First wash your hands.

2 'wash up'

In American English, if someone **washes up**, they wash parts of their body, especially their hands and face.

He went to the bathroom to wash up.

In British English, if you **wash up**, you wash all the dishes that have been used in cooking and eating a meal.

We washed up in the kitchen before having our coffee.

watch

→ see **see – look at – watch**

we

We is the subject of a verb. It is used in two main ways.

You can use **we** to talk about yourself together with someone else, but not the person you are speaking or writing to.

I shook his hand, and we both sat down.

You can also use **we** to include the person or people you are speaking or writing to.

Shall we have dinner together, Sally?

> Don't say 'you and we' or 'we and you'. Don't say '~~You and we must go and see John~~'. Say '**We** must go and see John'.

wear – in

1 'wear'

When you **wear** something, you have it on your body. You can **wear** clothes, shoes, a hat, gloves, jewellery, make-up, or a pair of glasses. The past tense form of **wear** is *wore*, not 'weared'. The past participle is *worn*.

He was wearing a brown shirt.
I've worn glasses all my life.

2 'in'

You can also use **in** to talk about what someone is wearing.

...a small girl in a blue dress.

> You don't usually use 'in' after 'be' when you are saying what someone is wearing. Don't say, for example, '~~Mary was in a red dress~~'. Say 'Mary **was wearing** a red dress'.

However, you can use **in** after **be** when you are also using a word such as **his** or **my**. You can say, for example, 'Mary was **in her red dress**'.

In is sometimes used to mean 'wearing only'. For example, 'George was **in** his underpants' means 'George was wearing only his underpants'.

He opened the door in his pyjamas.

weather – whether

1 'weather'

If you are talking about the **weather**, you are saying, for example, that it is raining, cloudy, sunny, hot, or cold.

The weather was good for the time of year.

> **Weather** is an uncountable noun. Don't use 'a' with it. Don't say, for example, '~~We can expect a bad weather tomorrow~~'. Say 'We can expect **bad weather** tomorrow'.

2 **'whether'**

Don't confuse **weather** with **whether**. Use **whether** when you are talking about a choice between two or more things.

I don't know whether to go out or stay at home.

wedding

→ see **marriage – wedding**

week

A **week** is a period of seven days.

She'll be back next week.

If something happens **in the week** or **during the week**, it happens on weekdays, rather than at the weekend.

In the week, we get up at seven.

→ see also **last**
→ see **next**

weekday

A **weekday** is any of the days of the week except Saturday or Sunday.

She spent every weekday at meetings.

You can say that something happens **on weekdays**.

I visited them on weekdays for lunch.

weekend

1 **'weekend'**

A **weekend** is a Saturday and the Sunday that comes after it.

Sometimes people include Friday evening as part of the weekend.

I spent the weekend at home.

2 regular events

British speakers say that something takes place **at weekends**.

The tower is open to the public at weekends.

American speakers usually say that something takes place **weekends** or **on weekends**.

On weekends I rarely do any work.
I stayed in the city weekends.

3 single events

You can use **the weekend** to talk about either the last weekend or the next weekend. You can use **at**, **during**, or **over** in front of **the weekend**.

Nine people were killed in road accidents at the weekend.
I'll call you over the weekend.

You can also use **this weekend** to talk about either the last weekend or the next weekend. Don't use any preposition in front of **this weekend**.

His first film was shown on television this weekend.
Let's go skiing this weekend.

well

1 used before an opinion

People sometimes say **well** when they are about to give their opinion. There is often no special reason for this, but sometimes **well** can show that someone is thinking.

'Is that right?' – 'Well, possibly.'

2 used as an adverb

Well is also an adverb that you use to say that something is done in a good way.

He did it well.

When **well** is an adverb, its comparative and superlative forms are *better* and *best*.

You play football better than I do.
I did best in physics in my class.

3 used as an adjective

Well is also an adjective. If you are **well**, you are healthy and not ill. You don't usually use **well** in front of a noun.

I am very well, thank you.

When **well** is an adjective, it does not have a comparative form. However, you can use **better** to say that someone is recovering, or has recovered, from an illness.

He seems better today.

→ see **better**

west

→ see topic: **North, South, East and West**

western

→ see topic: **North, South, East and West**

what

1 asking for information

You use **what** when you are asking for information about something.

What happened?
What did she say?
What is your name?

You can use **what** to ask for more information about a noun or noun phrase.

What qualifications do you have?
What car did you hire?

> Don't use 'what' when you are asking about one of a small number of people or things. For example, if someone has hurt their finger, don't say to them '~~What finger have you hurt?~~' Say '**Which** finger have you hurt?'
>
> *'Go down that road.' – 'Which one?'*

Use **what** when you are asking about the time.

What time is it?

2 giving an opinion or reaction

You can use **what** and a noun phrase to give an opinion about something when you are excited or angry.

What a marvellous idea!
What rubbish!

→ see also **how – what**

when

1 used in questions

You use **when** to ask about the time that something happened or will happen.

When did you arrive?
When are you getting married?

2 used to talk about a particular time

You use **when** to say that something happened, happens, or will happen at a particular time.

He left school when he was eleven.

If you are talking about the future, you use the present simple with **when**, not a future form.

Stop when you feel tired.

where

1 used in questions

You use **where** to ask questions about places.

Where's Jane?
Where does she live?
Where are you going?

2 used in reported questions

Where is often used in reported questions.

He asked me where we were.
I don't know where it is.

3 used in statements

You can use **where** to talk about a place you have already mentioned.

He came from Herne Bay, where he met Janine.

Where can also be used after a word such as **place**, **room** or **street**.

...the place where they work.
...the room where I did my homework.
...the street where my grandmother lives.

whether

→ see **weather – whether**

whole

1 'the whole of' and 'whole'

When you talk about **the whole of** something, you mean all of it.

...the whole of July.
...the whole of Europe.

Instead of using **the whole of** in front of a noun phrase beginning with a determiner such as **the**, **this** or **my**, you can simply use **whole**. For example, instead of saying 'The whole of the house was on fire', you can say '**The whole house** was on fire'.

They're the best in the whole world.
I've never told this to anyone else in my whole life.

You use **whole** to emphasize that you mean all of something.

I stayed there for a whole year.
There were whole chapters that I didn't understand.

> In front of plural nouns, **whole** does not have the same meaning as **all**. If you say '**All** the buildings have been destroyed', you mean that every building has been destroyed. If you say '**Whole** buildings have been destroyed', you mean that some buildings have been destroyed completely.

2 'on the whole'

You say **on the whole** to show that what you are saying is only true in general and may not be true in every case.

On the whole he liked Americans.

who's

→ see **whose – who's**

whose – who's

You use **whose** in questions when you are asking who something belongs to or is connected with.

Whose car were they in?
Whose is this?

Who is and **who has** are also sometimes pronounced /huːz/. When you write down what someone says, you can write 'who is' or 'who has' as **who's**. Don't write them as 'whose'.

'Edward drove me here.' – 'Who's Edward?'
...an American author who's moved to London.

why

1 used in questions

You use **why** when you are asking a question about the reason for something.

'I had to say no.' – 'Why?'
Why did you do it, Martin?

2 used for making suggestions

You can make a suggestion using **Why don't....?**

Why don't we all go?

3 used in reported questions

Why is often used in reported questions.

He wondered why she had come.

4 used in statements

You use **why** after the word **reason** in order to give an explanation.

There are several good reasons why I have a freezer.

will – shall

1 'will'

Will is used to make statements and ask questions about the future.

The concert will finish at 10.30pm.
When will you be home?

Will is often shortened to **'ll** and put after a name or a pronoun.

He'll come back.
Tom'll be here later.

The negative form of **will** is **will not**. This is often shortened to **won't** /wəunt/.

You won't hear much about it.

You can use **will you** to ask someone to do something.

Will you please be quiet?

Will is sometimes used to say that someone or something is able to do something.

This will cure anything.
The car won't start.

2 'shall'

You can use a question beginning with '**Shall we**...' to make a suggestion. 'Will we...' is not used in this way.

Shall we go and see a film?

win – defeat – beat

1 'win'

If you **win** a war, fight, game, or competition, you do better than the other person or people. The past tense form and past participle of **win** is *won* /wʌn/, not 'winned'.

The four local teams all won their games.

2 'defeat' and 'beat'

Don't say that someone 'wins' an enemy or opponent. In a war or battle, you say that one side **defeats** the other.

The French defeated the English army.

In a game or competition, you say that one person or side **defeats** or **beats** the other.

They were playing chess and she beat him.

winter

→ see topic: **Seasons**

wish

Wish is usually followed by a clause beginning with **that**. You can often leave out **that**. If you **wish** (that) something was true, you would like it to be true, although you know it is unlikely or impossible.

I wish (that) I lived nearer London.
I wish (that) I could paint.

If you **wish** that something **would** happen, you want it to happen, and you are angry, worried or disappointed because it has not happened already.

I wish he would come!
I wish you would try to understand.

Don't use 'wish' with a clause to say that you hope something good will happen to someone. Don't say, for example, '~~I wish you'll have a nice time in Finland~~'. Say 'I **hope you'll have** a nice time in Finland' or 'I **hope you have** a nice time in Finland'.

I hope I'll see you before you go.
I hope you like this village.

However, you can use **wish** followed by **you** and a noun phrase to say that you hope something good will happen to someone. For example, if you say 'I **wish you a happy birthday**', you mean 'I hope you have a happy birthday'.

I wish you both a good trip.

with

1 basic uses

If one person or thing is **with** another, they are together in one place.

I stayed with her for a while.

If you do something **with** a tool or object, you do it using that tool or object.

He pushed back his hair with his hand.

2 used to say who else is involved

You use **with** after verbs like **fight** or **argue**. For example, if two people are fighting, you can say that one person is fighting **with** the other.

He was always fighting with his brother.

Similarly, you can use **with** after nouns like **fight** or **argument**.

I had an argument with Greenberg.

3 used in descriptions

You can use **with** after a noun phrase to describe a thing or person.

...an old man with a beard.
...a house with three bedrooms.

You can use **with** like this to make it clear who or what you are talking about. For example, you can call someone 'the tall man **with** red hair'.

... the house with the blue door.

> Don't use 'with' to mention something that someone is wearing. Use **in**.
>
> *...an old woman in a black dress.*
>
> → see **wear – in**

woman – lady

You usually call an adult female person a **woman** /wumən/.

...a tall, dark-eyed woman in a brown dress.

The plural of **woman** is *women* /wɪmɪn/, not 'womans' or 'womens'.

There were men and women working in the fields.

You can use **lady** as a polite way of talking about a woman, especially if the woman is present.

There is a Japanese lady here, looking for the manager.

> It is more polite to call someone an **old lady** or an **elderly lady**, rather than an 'old woman'.
>
> *There's an old lady who rides a bike around town.*
> *...elderly ladies living on their own.*

If you are talking to a group of women, you call them **ladies**, not 'women'.

Ladies, could I have your attention, please?

work

1 used as a verb

If you **work**, you have a job and earn money for it.

I used to work in a hotel.

You can use **as** with **work** to say what a person's job is.

He worked as a teacher for 40 years.

The verb **work** has a different meaning in progressive forms than it does in simple forms. You use progressive forms to talk about a temporary job, but simple forms to talk about a permanent job. For example, if you say 'I'm working in London', you mean that you may soon move to a different place. If you say 'I work in London', you mean that London is your permanent place of work.

2 used as an uncountable noun

If you have **work**, you have a job and earn money for it.

...people who can't find work.

When someone does not have a job, you can say that they are **out of work**.

There are one and a half million people out of work in this country.

Work is also used to talk about the place where someone works.

He went to work by bus this morning.

worth

Worth can be a preposition or a noun.

1 used as a preposition

If something is **worth** an amount of money, you could sell it for that amount.

His yacht is worth $1.7 million.

> **Worth** is not a verb. Don't say '~~His yacht worths $1.7 million~~'.

2 used as a noun

You use **worth** as a noun after words like **pounds** or **dollars** to show how much money you would get for an amount of something.

...12 million pounds worth of gold and jewels.

> Don't talk about the 'worth' of something that someone owns. Don't say, for example, '~~The worth of his house has increased~~'. Say 'The **value** of his house has increased'.
>
> *What is the value of this painting?*

would

1 situations that are not real

You use **would** when you are talking about a situation that is not real.

If I had more money, I would buy a car.

2 requests, orders, and instructions

You can use **would** to make a request.

Would you do me a favour?

You can also use **would** to give an order or instruction.

Would you ask them to leave, please?

3 offers and invitations

You can say '**Would you...?**' when you are offering something to someone, or inviting them to do something.

Would you like a drink?
Would you prefer to stay with us?

4 being willing to do something

If someone said that they were willing to do something, you can say that they **would** do it. If they refused to do it, you can say that they **would not** do it or **wouldn't** do it.

He said he would help her.
She wouldn't talk to him.

5 short form

When **would** comes after a word like **I**, **he**, **you** or a name, you often write it as **'d**.

She said she'd come.

write

1 'write' and 'write down'

When you **write** something or **write** it **down**, you use a pen or pencil to make words, letters, or numbers. The past tense form of **write** is *wrote*. The past participle is *written*.

I wrote down what the boy said.

2 writing a letter

When you **write** a letter to someone, you write information or other things in a letter and send it to the person. You can say that you write someone something, or that you write something **to** someone.

I wrote him a long letter.
Once a week she wrote a letter to her husband.

If you **write to** someone, you write a letter to them.

She wrote to me last summer.

American speakers often leave out the 'to'.

If there is anything you want, write me.

You can write '**I am writing**...' at the beginning of a letter to say what you are writing about.

Dear Morris, I am writing to ask whether you will visit us this year.

> Don't write '~~I write to ask...~~'.

XYZ

year

A **year** is a period of twelve months, beginning on the first day of January and ending on the last day of December.

We had an election last year.

A **year** is also any period of twelve months.

The school has been empty for ten years.

You can use **year** when you are talking about the age of a person or thing.

She is now seventy-four years old.

When you use **year** to talk about age, you must use **old** after it. Don't say, for example, '~~She is now seventy-four years~~'.

→ see **old**

yes

You use **yes** to agree with someone or to say that something is true.

'Is that true?' – 'Yes.'

When someone asks a negative question, you must say **yes** if you want to give a positive answer. For example, if someone says 'Aren't you going out this evening?', say '**Yes**, I am'. Don't say '~~No, I am~~'.

'Didn't you get a dictionary from him?' – 'Yes, I did.'

Similarly, you say **yes** if you want to disagree with a negative statement. For example, if someone says 'He doesn't want to come', say '**Yes**, he does'. Don't say '~~No, he does~~'.

'That isn't true.' – 'Oh yes, it is.'

yesterday

Yesterday means the day before today.

It was hot yesterday.

You talk about the morning and afternoon of the day before today as **yesterday morning** and **yesterday afternoon**.

Heavy rain fell here yesterday afternoon.

You can also talk about **yesterday evening**, but it is more common to say **last night**.

I met your husband last night.

You can also use **last night** to talk about the previous night.

We left our bedroom window open last night.

Don't talk about '~~yesterday night~~'.

→ see also topic: **Times of the day**

yet

1 used in negative sentences

You use **yet** in negative sentences to say that something has not happened up to the present time, although it probably will happen. In conversation, you usually put **yet** at the end of a sentence.

It isn't dark yet.
I haven't decided yet.

In writing, you can put **yet** directly after **not**.

They have not yet set a date for the election.

2 used in questions

You often use **yet** in questions when you are asking if something has happened. You put **yet** at the end of the sentence.

Have you done that yet?

Some American speakers use the past simple in questions like these. They say, for example, '**Did** you **have** your lunch yet?'

3 'already'

Don't use 'yet' at the end of a question when you are surprised that something has happened sooner than expected. The word you use is **already**.

Is he here already?

→ see **already**

4 'still'

Don't use 'yet' to say that something is continuing to happen. For example, don't say '~~I am yet waiting for my luggage~~'. Say 'I am **still** waiting for my luggage.'

→ see **still**

you

→ see topic: **Talking about men and women**

you're

→ see **your – you're**

your – you're

1 'your'

You use **your** /jə/ or /jɔː/ to show that something belongs to the person or people that you are speaking to.

Where's your father?

2 'you're'

You are is also sometimes pronounced /jɔː/. You can write this as **you're**. Don't write it as 'your'.

You're quite right.

Adjectives that cannot be used in front of nouns

Some adjectives are not used in front of nouns. These adjectives are used after linking verbs such as **be, feel, look** or **seem**. For example, you can say '**She was alone**', but you cannot say '~~an alone girl~~'.

The following common adjectives are used only after linking verbs:

afraid	asleep
alike	awake
alive	glad
alone	hurt

For many of these adjectives there is another word you can use in front of a noun to express the same meaning.

Afraid must always go after a linking verb, but **frightened** can go in front of a noun or after a linking verb.

They seem afraid of you.
He was acting like a frightened child.

Alike must always go after a linking verb, but **similar** can go in front of a noun or after a linking verb.

They all looked alike to me.
They were given similar tasks.

Alive must always go after a linking verb, but **living** can go in front of a noun or after a linking verb.

I think his father is still alive.
I have no living relatives.

Asleep must always go after a linking verb, but **sleeping** can go in front of a noun or after a linking verb.

It was after midnight and she was asleep.
We both stared at the sleeping child.

Glad must always go after a linking verb, but **happy** or **cheerful** can go in front of a noun or after a linking verb.

I'm so glad you won.
She always seemed such a happy woman.

Hurt must always go after a linking verb, but **injured** can go in front of a noun or after a linking verb.

His friends asked him if he was hurt.
His injured leg was feeling better.

Times of the day

The main times of the day are:

morning
afternoon
evening
night

1 'in the' and 'on' with times of the day

You use **in the morning**, **in the afternoon**, **in the evening** and **in the night** to talk about when something happened.

She woke up in the morning feeling ill.
It rained in the night.

You also use **in the morning(s)**, **in the afternoon(s)** and **in the evening(s)** to talk about things that happen regularly at these times.

You could sit there in the evening and listen to the radio.
Most of us play golf in the afternoons.

Don't say 'in the night' or 'in the nights' for things that happen regularly. Say **at night**.

We lock the doors at night.

When you say the name of the day in front of **morning**, **afternoon**, **evening** and **night**, you use **on**.

Mum arrived at my house on Friday evening.
We go swimming on Saturday mornings.

2 'this' with times of the day

You use **this** in front of **morning**, **afternoon** and **evening** to talk about times in the present day.

The report will be on your desk this afternoon.

> Don't say ~~this night~~ or ~~this day~~. Say **tonight** and **today**.
>
> *I'll be late for dinner tonight.*
> *I had lunch with Delia today.*

3 'yesterday' with times of the day

You use **yesterday** in front of **morning**, **afternoon** and **evening** to talk about times in the previous day.

He went to work yesterday morning as usual.

> Don't say ~~yesterday night~~. Say **last night**.
>
> *Where were you last night?*

4 'tomorrow' with times of the day

You use **tomorrow** in front of **morning**, **afternoon**, **evening** and **night** to talk about times in the day after the present day.

Would you like to come for dinner tomorrow night?

Seasons

The seasons are:

spring summer autumn winter

Speakers of American English say **fall** instead of 'autumn'.

To say that something happens in a particular season, you use **in spring**, **in summer** etc., or **in the spring**, **in the summer** etc. You use these both when something happens once and when it happens every year.

Adam took his exams in the spring.
This house is too hot in summer and freezing in winter.

> Even if something happens every year, you still use the singular form of **spring**, **summer**, **autumn** and **winter**. Don't say '~~The house is too hot in summers.~~'

Transport

When you say what kind of transport you use to travel somewhere, you use **by** in front of the name of the vehicle.

She decided to travel by train.
I go to work by car.

> If you walk somewhere, don't say that you go '~~by foot~~'. Say that you go **on foot**.
>
> *I decided to continue my journey on foot.*

When you are talking about public transport, you can use the verbs **take** and **catch** instead of 'go by' or 'travel by'.

He left London and took a train to Leeds.
We caught the bus into the town centre.

When you are talking about entering or leaving a vehicle, you usually use **get on** and **get off**. You say that you **get on** and **get off** a bicycle, train, bus, boat or plane.

I was happy to get on the plane and relax for an hour.
I got off the train at Liverpool.

However, you use **get into** and **get out of** when you are talking about entering or leaving a car or taxi.

They got into a car and drove away.
A thin man got out of the taxi and approached me.

Meals

1 'breakfast'

Breakfast is the first meal of the day. You eat it in the morning, just after you get up.

I open the mail immediately after breakfast.

2 'lunch' and 'dinner'

Lunch is the meal eaten in the middle of the day.

Where did you have lunch?

Dinner is the main meal of the day, usually eaten in the evening.

They invited us for dinner at their house.

3 'tea' and 'supper'

Tea is a small meal eaten in the afternoon, especially in hotels and cafés, usually with sandwiches and cakes.

I bought some chocolate biscuits for tea.

Some people call the meal they have in the evening **tea**, especially if it is not their main meal, or if it is a meal for children.

Katie's friend stayed for tea.

Supper is a meal you eat in the evening. Some people call their main meal **supper**, while others use it for a small meal that they eat just before going to bed.

We had eggs and toast for supper.

4 'have'

You often use **have** to say that someone eats a meal.

I haven't had breakfast yet.

5 'make'

When someone prepares a meal, you can say that they **make** it.
Stella made dinner for all of us.

6 'at'

You use **at** in front of the name of a meal to talk about something that happens while you are eating that meal.

They spoke about it at dinner that night.

7 'for'

You use **for** after the name of a food and in front of the name of a meal to say what someone eats at that meal.

He had a sandwich for lunch.

You also use **for** in front of the name of a meal to talk about going somewhere to eat or inviting someone to eat with you.

Shall we go out somewhere for lunch?
I invited her to join us for dinner that night.

8 'my', 'your', 'his', 'her', 'our' and 'their'

You don't usually use 'the' in front of names of meals. Don't say, for example '~~I haven't had the breakfast yet~~' or '~~He had a sandwich for the lunch~~'. However, you can use words like **my**, **his** and **her** in front of names of meals.

Go and have your breakfast while I read these letters.
I went downstairs to make my lunch.

Places

1 'the'

When you talk in a general way about going to a place such as a school, church or hospital, don't use 'the' or 'a'. Don't say, for example '~~I take my children to the school by car.~~' Say 'I take my children **to school** by car.'

He was sent to prison for a very long time.
Are you going to church tomorrow?

Mosque and **synagogue** are not used like **church**. For them, you do use **the**.

I encourage my children to go to the mosque and pray.

You use **the** after a preposition when you are talking about a particular place you have already mentioned, or when you add a phrase to make it clear which place you are talking about.

We drove to the hospital in silence.
She wanted to go to the school in Paris.

2 'at' and 'in'

If you want to say where someone is, without mentioning the specific place, you use **at** and **in** without 'the'. With **school, college** and **university**, British speakers usually use **at**.

I was at university with her.

Speakers of American English usually use **in** with these words.

They met in high school.

With **prison** and **hospital**, you use **in**.

He died in prison.

Negatives

1 'never'

You use **never** to say that something did not, does not, or will not happen at any time.

Never comes in front of the verb, unless the verb is **be**, in which case it goes after it.

They never take risks.
The road by the river was never quiet.

Never comes after an auxiliary verb but before the next verb.

My husband says he will never retire.
I have never seen anything like this before.

2 'none'

None means 'not one' or 'not any'.

None of is used in front of a plural noun phrase or in front of an uncountable noun.

None of our players got through to the finals.
Their employers cover none of the cost of this training.

None can also be used on its own as a pronoun.

I went to get a biscuit, but there were none left.

3 'no-one'

No-one means 'not a single person'. It is followed by a singular verb.

No-one knows where he is.

4 'not'

Not is used with verbs to form negative sentences. It is often shortened to **n't**.

You put **not** after the first auxiliary verb, if there is one.

They might not notice.

> You almost always need an auxiliary verb in front of **not**. If there is no other auxiliary, you use **do**. Don't say, for example, '~~I not liked it~~'. Say 'I **didn't like** it'.

However, when you use **not** with **be**, don't use an auxiliary verb. Simply put **not** after **be**.

The program was not a success.

5 'nothing'

Nothing means 'not a single thing'. It is followed by a singular verb.

Nothing is happening.

6 'nowhere'

You use **nowhere** to say that there is no place where something happens or can happen.

There was nowhere to hide.

7 two negatives

> Don't use two negative words in the same sentence.
>
> If you use a negative word like **none** or **nowhere**, don't make the verb negative with 'not' as well. Don't say, for example '~~None of the children weren't ready.~~' Say 'None of the children were ready.'

> If you make a verb negative with 'not', you must use a positive word with it. Don't say, for example, '~~I don't know nothing about it.~~' Say 'I **don't know anything** about it'.

North, South, East and West

1 basic meanings

North is the direction that is to your left when you are looking towards the place where the sun comes up.

Our route went from north to south.

South is the direction that is to your right when you are looking towards the place where the sun comes up.

We have good relations with our neighbours in the south.

East is the direction that you look towards in order to see the sun come up.

There are huge mountains to the east.

West is the direction that you look towards in order to see the sun go down.

They come from the west of Ireland.

2 with 'the'

You use **the** in front of **north**, **south**, **east** and **west** to talk about a part of a country or region.

Their troops were based in the north.
We were on holiday in the south of France.

3 used as adjectives

North, **south**, **east** and **west** are used as adjectives to mean 'facing a particular direction' or 'in a particular area'. They are often used with names of towns and cities, or in place names.

We sailed round the north coast of Australia.
She bought a house in east London.
They moved to South Carolina.

> You don't usually use 'north', 'south', 'east' and 'west' in front of names of countries or continents. Instead, you say **northern**, **southern**, **eastern** and **western**. Don't say, for example, ~~Most of the students come from west Europe.~~. Say 'Most of the students come from western Europe'.
>
> You don't usually use 'north', 'south', 'east' and 'west' in front of words like 'part', 'area', 'region', 'town' or 'city'. Don't say, for example, '~~There is a lot of industry in the north cities.~~' Say 'There is a lot of industry in the northern cities.'

You use **north**, **south**, **east** and **west** in front of **wind** to say where the wind is blowing from.

The east wind had increased throughout the day.

4 used as adverbs

North, **south**, **east** and **west** are used as adverbs to mean 'towards a particular direction'.

I was travelling north from Florida to Alabama.

Talking about men and women

1 'they', 'them' and 'their' to refer to a single person

When you are talking about a person and you do not know if the person is male or female, you use the words **they**, **them** and **their**. You use these to refer back to words and phrases like **anyone**, **someone**, **each person** and **a person**.

Ask an adult if they can help you.

Anyone with a raincoat should take it with them.

2 'they', 'them' and 'their' to refer to a group of people

You also use the words **they**, **them** and **their** with **everyone**, **everybody**, **every person** and so on.

Everyone was dressed in their best clothes.

In very formal writing, you can use **he or she**, **him or her** or **his or her** instead.

Each patient can make choices about his or her treatment.

3 'you' and 'one'

You is often used to refer to people in general. **You** is followed by a plural form of a verb.

To be a good doctor you need to have good communication skills.
It's easier to solve a problem when you know a lot about it.

One can be used in the same way, but it is much more formal. **One** is followed by a singular form of a verb.

One doesn't talk about politics at the club.

4 'man' and 'mankind'

Man and **mankind** are sometimes used to refer to people in general. For example, instead of saying 'Human beings are destroying the environment', you can say '**Man** is destroying the environment' or '**Mankind** is destroying the environment'. You use a singular form of a verb.

...the most dangerous substance known to man.
Mankind depends on trees.

Many people avoid using **man** and **mankind** like this because it makes it seem as though men are more important than women. Instead, they use **people** or **humans**.

Where you put adverbs

1 adverbs of manner, place and time

Adverbs of manner, place and time are usually used after the main verb in a sentence.

They looked anxiously at each other.
She is sleeping upstairs.
They will leave tomorrow.

If the verb has an object, the adverb comes after the object.

I read the instructions carefully.
We will see you tomorrow.

Adverbs of time can also go at the beginning of a sentence.

Later, the same men were seen in the High Street.

2 adverbs of possibility and frequency

Some adverbs come in front of the main verb, unless the verb is **be**. These include adverbs of possibility (such as **probably**, **definitely** and **certainly**), adverbs of frequency (such as **often**, **sometimes** and **rarely**) and the adverb **still**.

Marcus definitely wants to come.
She sometimes visits us.
She still lives in London.

If the verb is **be**, you put the adverb after it.

They are probably expensive.

She was still beautiful.

Where there is an auxiliary verb in the sentence, the adverb comes after it.

They will probably leave.
They have often helped us in the past.
He could still get into serious trouble.

3 'only'

Only means 'not anyone or anything else' or 'not more than a particular size or amount'.

Only is used in front of the subject of a sentence.

Only his close friends knew how ill he was.

Only is also used in front of a verb or after the first auxiliary verb, unless the verb is **be**.

They only went as far as the station.
We could only choose two of them.

If the verb is **be**, you put **only** after it.

There is only one train that goes from Denmark to Sweden by night.